The Conversation on Water

Critical Conversations

Martin LaMonica, Series Editor

The Conversation U.S. is an independent, nonprofit news organization dedicated to delivering expert knowledge to the public through journalism. Every day The Conversation produces 10–12 stories through a collaboration between scholars and editors, with the scholars writing explanatory journalism and analysis based on their research and the editors helping them translate it into plain language. The articles can be read on TheConversation.com and have been republished by more than a thousand newspapers and websites through a Creative Commons license, meaning that the content is always free to read and republish.

The book series Critical Conversations is published collaboratively by The Conversation U.S. and Johns Hopkins University Press. Each volume in the series features a curated selection of subject-specific articles from The Conversation and is guest-edited by an expert scholar of the subject.

■

The Conversation on Water is guest-edited by Andrea K. Gerlak, Professor in the School of Geography, Development and Environment and Director of the Udall Center for Studies in Public Policy at the University of Arizona.

Martin LaMonica is Executive Editor and Project Manager.
Kira Barrett is Editorial Assistant.
Gita Zimmerman is Illustrator of the part-title images, overseen by Conversation Marketing and Communications Manager **Anissa Cooke-Batista**.
Beth Daley is Editor and General Manager of The Conversation U.S.
Bruce Wilson is Chief Innovation and Development Officer of The Conversation U.S.

We would like to express our gratitude to the editors and scholars who produced the articles collected here and to thank our colleagues and funders who allow us to do this important work in the public interest.

THE C⬭NVERSATION
on Water

edited by Andrea K. Gerlak

Johns Hopkins University Press

BALTIMORE

Johns Hopkins University Press
2715 North Charles Street
Baltimore, Maryland 21218
www.press.jhu.edu

Library of Congress Cataloging-in-Publication Data

Names: Gerlak, Andrea Kristen, editor.
Title: The conversation on water / edited by Andrea K. Gerlak.
Description: Baltimore : Johns Hopkins University Press, 2023. | Series:
  Critical conversations | Includes bibliographical references and index.
Identifiers: LCCN 2022030508 | ISBN 9781421446202 (paperback) | ISBN
  9781421446219 (ebook)
Subjects: LCSH: Water supply. | Water security. | Water—Pollution. |
  BISAC: SCIENCE / Environmental Science (see also Chemistry /
  Environmental) | POLITICAL SCIENCE / Public Policy / Environmental
  Policy
Classification: LCC TD355 .C66 2023 | DDC 363.6/1—dc23/eng/20221012
LC record available at https://lccn.loc.gov/2022030508

A catalog record for this book is available from the British Library.

Special discounts are available for bulk purchases of this book.
For more information, please contact Special Sales at specialsales@jh.edu.

Contents

Part IV.
The Lifeblood of Human Society   144

Part V.
Preserving Our Oceans   176

First, a Quick Story

IN THE FALL OF 2014, my wife sent me a listing for an environment and energy editor job at a place called The Conversation. I had vaguely heard of this outfit, one of many startups trying to innovate within the troubled world of journalism. As a longtime reporter and editor, I had little optimism that anyone could fix the media's broken business model. But I was intrigued by the organization's approach of having college professors and researchers fill the gap left by layoff after layoff of experienced reporters. I emailed the managing editor and landed an interview.

Days after I left the interview held in a cramped basement office on the Boston University campus, the mission underpinning the venture—to improve the public discourse—stayed with me. Could academics, working with journalists, help fulfill the vital role of journalism to inform the public by sharing more facts and knowledge?

As I now write this foreword years later, I can say that the power of The Conversation's founding idea and novel editorial model endures. And millions of people benefit every day. A media nonprofit with editions in multiple countries, we publish daily news analysis and explanatory journalism that comes out of a collaboration between academics and journalists. Put another way, it's like a digital newspaper whose reporters are researchers and professors with deep subject expertise and whose editors are journalists who themselves are topic specialists with years of experience covering the news.

We have a website, multiple email newsletters, and many outposts on social media. Why make a book from our daily journalism? And who would want to read it?

As an independent media nonprofit, we exist to serve the public with accurate and reliable information you can use to navigate an increasingly complex world. We are funded by universities, foundations, and individual donors. This collection's contributions are grounded in academic research, since the authors—sometimes people who have worked in their field for decades—are writing about their areas of expertise. There are citations to peer-reviewed articles and books throughout, and this collection itself has gone through peer review.

But this rich information is also accessible. Working with journalist-editors, the academics write with a general reader in mind—anyone who is curious to learn more about a subject and values the knowledge that comes from years of study and academic achievement. The format is what you'd expect to read in most media outlets as well: stories of somewhere between 800 and 1,200 words long that can be read in a brief amount of time. Since many of the chapters in this collection were originally written in response to events in the news, most have been modified slightly to remove references that would date them. Inevitably, new developments or events will supersede what we published previously, but we have tried to ensure the chapters are accurate.

The themes we have chosen for these collections reflect our editorial values and the makeup of our newsroom. At The Conversation, we have the luxury of having dedicated editors in a wide range of subjects, from education to climate change—something that few media outlets still have the resources to do.

Perhaps the biggest payoff from this multidisciplinary approach comes when we examine one subject in depth, as we do in this book. It's enlightening, and often quite fun, to discover what you didn't know by reading an essay from an ethicist, for instance, a few pages after hearing what a microbiologist or a historian has to say. Having multiple entry points into a subject helps us connect the dots to see the bigger picture. Also, solving thorny societal problems requires a wide-angle view: discussions about new technologies and scientific developments, for example, cannot be separated from the societal impacts they have, just as, say, dealing with environmental challenges requires input from the physical sciences, politics, sociology, and more.

It's worth noting that the multiple voices and different viewpoints in this collection are just that—varied. That means the tone of chapters will vary, and some authors may not even agree on certain points. But that's OK, because the goal is to give you, the reader, the context and a foundational understanding of issues that are important to living in today's society.

Our hope is that, after reading this collection, you will feel better equipped to make sense of news in the headlines or to grasp the significance of new research. In other cases, you may just be entertained by a good story.

Martin LaMonica

The Not-So-Hidden Water Crisis

WE TAKE WATER FOR GRANTED in the United States. We expect it to flow freely from our taps—clean and at low cost. Historically, the United States has been home to some of the most reliable and safest drinking water in the world. It is easy to get complacent and even out of touch with our water, unaware of its source. Catastrophe can seem isolated to natural disasters like hurricanes and winter cold snaps that only temporarily disrupt water infrastructure and availability. This is in sharp contrast to other parts of the world facing a chronic lack of access to safe and affordable water and sanitation.

But even in places in the United States that have sound drinking water systems, new threats continue to emerge. And for millions of people in this country, water contamination or unavailability is a serious problem, the result of long-standing inequities in our society. As the spectrum of water-related concerns grows, Americans will need not only to reckon with decades of underinvestment but also to recognize the many social justice issues intertwined with this vital yet often under-appreciated natural resource.

Cracks in the System Are Appearing

Flint, Michigan. Over recent years, Americans have come to know this city in Michigan just outside Detroit as ground zero for water contamination. In 2014, when the city switched its drinking water supply from Detroit's system to the Flint River in an effort to save money, inadequate testing and treatment of the water resulted in major problems for public health and water quality for local residents. Despite their protests and concerns over the smell and color of the water, and as lead leached from aging pipes into homes, community members, in this majority-Black city where 40 percent of people live in poverty, were simply ignored by city and state officials. A series of unfortunate events unfolded including water-boiling initiatives, leaked emails, university-led water tests, resignations, lawsuits, and, ultimately, manslaughter charges. Flint is a painful object lesson in the consequences of deferring critical investments in water infrastructure. But even more than that, Flint reveals inequities in access to clean water, which so many communities of color fail to enjoy in the United States today.[1]

The Flint, Michigan, water crisis raised national concerns about the management of our water systems and the overall safety of drinking water. But the story doesn't end in Flint.

As Justin Worland reported in *Time* magazine, "In Michigan, officials put an entire community at risk to save money, then lost a bet that the risks would go unnoticed. Similar wagers are placed by politicians and policymakers across the country."[2] Today, many communities in the United States—from Pittsburgh to Newark, and Milwaukee to Tacoma—are facing similar challenges with lead contamination in water supplies. Estimates suggest that some 18 million people are served each year by drinking water systems with lead violations.[3]

And lead isn't our only concern. Residents of Las Vegas have been forced to deal with perchlorate, a chemical found in rocket fuel and used by nearby manufacturers, which has been leaching into Lake Mead through groundwater. Appalachian mining communities drink water contaminated by sewage and coal slurry spills containing dangerous heavy metals. Water supplies in farming communities in America's heartland often exceed legal limits for nitrates, caused by nitrogen-based fertilizers used for farming. Some communities, like the Navajo and Hopi Nations in the southwestern United States, experience groundwater contamination due to elevated levels of arsenic and radioactive industrial waste from mining.

I've been studying the politics of water for almost three decades now as a university professor and researcher. Over the course of my career, I have seen a significant growth in knowledge around how we manage water. We are coming to better understand the scope of the problem of mismanagement, and it is not as isolated and limited geographically as one might expect. In 2017 the Natural Resources Defense Council released a report indicating that nearly 77 million Americans, or almost a quarter of Americans, were served by water systems reporting violations of the Safe Drinking Water Act, the main federal law regulating contaminants in drinking water.[4] Violations ranged from contamination with arsenic and nitrates to failures in testing and reporting contamination levels. Small systems in rural areas account for nearly 70 percent of all violations, suggesting that rural Americans could be at the greatest risk from some drinking water contaminants. A 2022 report by the Environmental Integrity Project found that the US Environmental Protection Agency had failed to update its technology-based limits for water pollution control systems used by industries in over three decades, allowing for more pollution to enter our waterways from oil refineries, chemical plants, other industries.[5]

The American Society of Engineers consistently gives America's water infrastructure inferior grades. In 2021, our drinking water infrastructure was given a C grade.[6] Our aging and underfunded systems are characterized by water leaks responsible for the loss of some 6 billion gallons of water every day. Cities from Atlanta to Cleveland to Pittsburgh are losing upward of 30 percent of their water each year to leaking pipes.

But in many communities, the problems are much more than a lack of federal enforcement. In a recent study on urban water access in the United States, Meehan and coauthors found that some 471,000 households, or 1.1 million people, lacked a piped water connection.[7] The US Water Alliance reported that some two million Americans live without running water and basic indoor plumbing.[8] Some 30–40 percent of homes in the Navajo and Hopi Nations in the southwestern US have no sewer connection, and 18 percent have no running water, with residents being forced to rely on tanker truck deliveries or the purchase of bottled water.[9] When it comes to water infrastructure, America's challenges increasingly resemble those of a developing country.

The list of known health problems related to poor water quality continues to grow. We know for certain that water pollution kills. A study reported in the *Lancet* found pollution to be the largest environmental cause of disease and death in the world today, responsible for an estimated nine million premature deaths.[10] Toxins in water can cause a variety of health issues, including cancer, hormone disruption, fertility problems, and altered brain and nervous system functions. We are just beginning to understand the negative human health impacts of some pollutants, such as microplastics, that we consume in our tap water and bottled water.[11]

Our children are especially vulnerable. Children often have greater exposure to toxic pollutants in drinking water because they drink more water per pound of body weight, and they are more vulnerable to pollutants since their bodies are still growing and chemicals cause greater harm to developing organs and tissues.[12] Many Americans are increasingly expressing concern over the quality of tap water in their children's schools. A recent poll found that 45 percent of parents with school-aged children under the age of 18 questioned the safety of the tap water in their children's schools.[13]

Marginalized Communities Have It the Hardest

Some households simply cannot afford to keep water flowing. While Americans have historically enjoyed relatively low water payment rates, this is changing. Food and Water Watch identified a "secret water crisis," reporting that household water bills have increased at three times the rate of inflation, even though income has fallen in real terms.[14] The lowest 20 percent of income earners spend nearly one-fifth of their income on water.[15] According to a study from Michigan State University, water is unaffordable for 1 out of 10 households.[16] Research over the past decade found that across a number of US cities, water bills increased an average of 80 percent, with more than two-fifths of residents in some cities living in neighborhoods with unaffordable bills.[17]

The COVID-19 pandemic exposed water affordability issues in the United States.[18] The lack of affordable access to water is all the more devastating given the extra sanitation needs during the pandemic.[19] Some reporting found that since the beginning of the pandemic, many people fell behind on water bills and, as a result, suffered increased shutoffs owing to nonpayment.[20] The 2020 Coronavirus Aid, Relief, and Economic Security Act provided some relief for debt from household water bills, and many state utilities temporarily suspended water shutoffs and eliminated late fees.[21] Yet a recent report showed that water debt is 10 times higher in Chicago's Black neighborhoods than it is in the city's white neighborhoods.[22] COVID also hit the water utilities hard. It is estimated that water utilities will experience a negative aggregate financial impact of somewhere between US$13 and US$16 billion because of revenue losses and increased operational costs during the pandemic.[23]

Water and sanitation problems are especially bad in communities of color and for disenfranchised Americans. Drinking water systems with regular violations of federal safety standards are 40 percent more likely to occur in places with higher percentages of residents of color.[24] Native Americans are 19 times more likely to lack indoor plumbing than their white counterparts, and African American and Latinx households lack indoor plumbing at almost twice the rate of white households.[25] African American, Native American, and other non-white households are disproportionately affected by utility disconnections, and Black households are twice as likely to be disconnected as those of whites.[26] Meehan and coauthors found that, in urban settings, houses without plumbing were more likely to be headed by people of color and to be poorer.[27]

How Did We Get Here?

We did not get here overnight; rather, a series of factors help to explain the emerging water crisis in the United States. Shrinking budgets, aging infrastructure, emerging contaminants, and increasing pressure from climate change all threaten our access to clean, reliable water in the United States.

Funding for water infrastructure has been on the decline since the 1970s. We have simply not invested in the necessary ways to maintain and upgrade our water systems. Most of our water infrastructure was built 50 or more years ago in the post–World War II era, and in some older urban areas in the United States, systems are more than a century old. The United States has been falling in global rankings of infrastructure quality, lagging behind other developed nations.[28]

The federal government's share of capital spending in the water sector fell from 63 percent in 1977 to just 9 percent of total capital spending in 2017.[29] States and local governments have been filling the gap, providing roughly two-thirds of public spending for capital investment in water infrastructure since the 1980s.[30] The lack of federal investment has forced local water systems to raise rates to cover their costs, which helps to account for the rising water bills and growing unaffordability of drinking water in many parts of the United States.

Small water systems characterized by meager tax bases, low-income communities, and often small populations have been hit the hardest.[31] These systems struggle to attract investment and often then have no option but to try to fund repairs and maintenance by raising rates.[32] Furthermore, many small utilities in rural communities lack the ability to do appropriate water testing, or they don't have the technical expertise to design improvement projects or even to complete the necessary technical documents to apply for federal funding.[33]

Reinvesting in our aging water infrastructure will not be inexpensive. The projections are mind-boggling. The US Environmental Protection Agency estimates that we will need to spend US$632 billion over the next decade on water infrastructure.[34] The costs of replacing infrastructure in rural communities alone may amount to almost US$190 billion in the coming decades.[35] According to the American Water Works Association, the investment needed for buried drinking water infrastructure will total more than US$1 trillion nationwide over the next 25 years.[36] Postponing investment will only make the problem worse by increasing the odds of facing the high costs associated with broken water mains and other infrastructure failures.

In addition to shrinking budgets and aging infrastructure, new emerging pollutants like polyfluoroalkyl and perfluoroalkyl substances—commonly referred to as PFAS—challenge our water systems. A Harvard University study found unsafe levels of PFAS in the drinking water of 33 states, affecting some six million Americans.[37] PFAS have been detected in drinking water supplies in many cities across the country, including Atlanta, Boston, New York, Chicago, Miami, and Washington, DC. PFAS are toxic chemicals found in everyday products like waterproof jackets, nonstick pans, cosmetics, and food packaging, and they are used in firefighting foam on military bases and at commercial airports. Scientists call PFAS "forever chemicals" because their chemistry keeps them from breaking down under typical environmental conditions.[38]

Scientists increasingly see links between the contaminants and liver damage, weakened immune systems, and cancer. According to Rick Rediske, a professor at the Annis Water Resources Institute in Michigan, "DDT and pesticides go into our fat. Lead goes into our bones. Mercury goes into muscle. Because PFAS are carried around in our blood and aren't discarded, they naturally concentrate over time. And they attach to the proteins that carry antibodies, cholesterol and hormones, that's why you get so many different health effects caused by these compounds."[39] We are only just getting a handle on the extent of this contamination, and regulation and treatment technologies have been slow in coming and expensive.[40] In part, this is because we can't keep up with their production, and we have failed to ban them as Europe has.[41] For decades, the US Environmental Protection Agency has failed to regulate PFAS,[42] and only recently has the Biden administration initiated efforts to address these challenges.[43]

In Tucson, Arizona, where I live, city officials recently closed down a major water treatment facility, which delivered drinking water to 60,000 residents, because of PFAS contamination that threatened to overwhelm groundwater filtration systems.[44] Ironically, the treatment facility that shut down is nearby the site of a water contamination crisis where the community was exposed to trichloroethylene over decades, resulting in a variety of illnesses and cancers.[45] The earlier trichloroethylene contamination resulted in lawsuits and settlements, the designation of a Superfund site, the closing of contaminated wells, a massive cleanup effort, and a sense of mistrust in government officials.[46] The closing of wells puts strains on an already-overburdened water supply in a time of climate change and long-term drought conditions regionally.

Finally, we can expect that increased pressure from climate change will make matters worse. Scientists predict that climate change will produce changes in stream flows and increased storm surges, which could negatively impact our water resources. Coastal flooding, heat waves, and drought can be expected.[47] Consider the drought in the American Southwest that has made national headlines in recent years.[48] Today some 40 million people in the southwestern corner of the country who have come to rely on the Colorado River are experiencing significant declines in flow as a result of a 20-year drought and climate change.[49] Images of the bathtub ring around Lake Mead, a major reservoir that stores water for the region, are a stark reminder of the impacts of climate change and the shortages ahead.[50] Generally, as water sources decline, the concentration of contaminants increases, causing interactions between water quantity and quality. Polluted water generally means less available water.

Water around the World

The United States is not alone when it comes to these water challenges. Water demand is on the rise worldwide, as a result of population growth and socioeconomic development. Today, over 2 billion people live in countries experiencing high water stress, and about 4 billion people experience severe water scarcity during at least one month of the year.[51] A joint report of the World Health Organization and the United Nations Children's Fund stated that some 2.2 billion people around the world do not have safely managed drinking water, that 4.2 billion people do not have safely managed sanitation, and that 3 billion lack basic handwashing facilities.[52] Experts expect that global water demand will increase by 20–30 percent over the next three decades in response to domestic and industrial demand.[53] Climate change is expected to exacerbate shortages in already-water-stressed regions of the world and increase the number of these regions.[54]

Similar to the United States, many countries are experiencing water infrastructure challenges. A recent United Nations report detailed chronic underinvestment in water infrastructure around the world.[55] The coronavirus pandemic has helped to shed light on these issues, as communities around the world, especially in Africa and Asia, found themselves more vulnerable to the virus when they lacked access to clean water and sanitation. A real challenge is that investments in water have been seen narrowly as social or environmental in nature. But there are, rather, real economic benefits to improving access to water and sanitation. The return on investment can be quite high, with a global average benefit-cost ratio of 2.0 for improved drinking water and 5.5 for improved sanitation.[56]

Globally, great strides have been made in recent years to address challenges for access and sanitation. Development efforts over the past two decades have significantly improved access to better drinking water and improved sanitation in many parts of the Global South.[57] The United Nations has come to recognize that access to safe water and proper sanitation is a human right, and development organizations, called nongovernmental organizations, and private sector actors have been making progress in expanding access to clean water and sanitation.[58] The United Nations' Sustainable Development Goals, which guide development efforts globally, put water front-and-center, calling for the "availability and sustainable management of water and sanitation for all."[59]

But challenges and notable inequities remain. Indigenous peoples suffer disproportionate violations of their right to safe drinking water and sanitation.[60] This has included loss of local control of ancestral lands and water, as well as exclusion from decision-making about water.[61] Children are especially vulnerable. Every day, children around the world become ill and many die from preventable water- and sanitation-related diseases. Climate change poses significant risks to children's health by diminishing access to safe water and sanitation.[62] Affordability is a challenge for many communities, especially in the Global South. Households without access to reliable, clean, piped public water often pay excessive amounts for water from alternative informal providers of water.[63]

Even so, many communities around the world are responding with novel approaches to water management that conserve resources and engage their citizens. In South Africa, the pandemic helped to galvanize municipal investment in water and sanitation.[64] With new technologies and innovative infrastructure, cities like Singapore are moving toward water independence by recycling water on a massive scale.[65]

In China, nongovernmental organizations and local citizens are helping to monitor the water quality of urban waterways, thereby assisting in decreasing pollution in their communities.[66] First Nations in Australia's Murray–Darling Basin are making progress in their fight to reclaim their historical water allocations. We can learn a good deal from how other communities are realizing commitments around water by their investing in infrastructure, tackling public health challenges, and promoting greater access and affordability.

Pathways Forward

The challenges concerning water are multifaceted and interlinked, demanding thoughtful and collaborative problem solving. Consider again the Flint water crisis. From one perspective it can be viewed as the fault of water treatment, regulation, and infrastructure, but it is also a story of political and democratic failure, the result of historical factors like structural racism, deindustrialization, and depopulation.[67] Even so, if examined in a more positive light, it can be seen as a powerful example of community mobilization that called government actions to account and brought a voice to an otherwise disenfranchised community.[68]

The good news is that we are gaining a better understanding of the many dimensions of the water crisis in the United States, and a growing chorus of academics, practitioners, and activists are working hard to identify, analyze, and address this crisis.[69] The pathways forward will likely need to include innovative solutions and technology, as well as new governance mechanisms that reflect public trust and acceptance.[70] We can learn from other communities around the world facing similar challenges.

Some communities are facing up to their water problems and experimenting with innovative strategies to build water resilience. Artificial intelligence, machine learning, high-tech listening devices—and even small robots that can travel in our water pipes—are being experimented with in cities around the globe to better pinpoint water leaks and target places for pipe replacement.[71] Breweries from Canada to Sweden, and across the American West, are using treated wastewater to make their product, thus helping to reduce demand in what is otherwise a water-intensive manufacturing process.[72] Many cities in the United States are adopting drought planning, expanding conservation and landscaping programs, and using water sensors to provide homeowners with data on water use in an effort to reduce demand.[73] There is also a growing number of communities adopting natural solutions like green infrastructure to realize the benefits of mitigating flooding for improved water quality and providing more urban green space to residents.[74] Communities are inaugurating innovative incentive programs and needed regulation and are recognizing how such strategies can redress past inequities in access.[75]

One sure bet we can make is that we will need a long-term investment plan to rebuild and repair our water infrastructure in order to respond to climate change and water quality concerns. Modernizing our drinking water infrastructure will require repairing old pipes and upgrading wastewater treatment facilities. More robust testing for drinking water contaminants is needed as well. Furthermore, we will need to strengthen enforcement of drinking water regulations and make new ones to protect human health.[76]

Increasingly, Americans are coming to understand the challenges for clean water supply that lie ahead. Today, two out of three Americans believe their community is vulnerable to a water crisis. A majority of Americans believe that immediate and significant investments in water infrastructure are necessary.[77] Studies conducted by water utilities, engineers, and advocacy groups estimate that the needed investments in US water infrastructure could add US$220 billion and 1.3 million jobs to the nation's economy annually.[78]

The infrastructure package passed by Congress in 2021 promised US$55 billion to expand access to clean drinking water for small water systems and underserved communities, to replace lead pipelines, and to control other contaminants.[79] The legislation dedicated significant funding that can be used for drinking water projects in tribal communities. Some US$3.5 billion has been allocated for the federal Indian Health Service, which amounts to the largest single infusion of money into Indian Country.[80] But the legislation is just a first step. To ensure long-term sustainability and resilience in tribal drinking water systems, the federal government needs to make its existing programs more accessible to tribes, improve coordination across the multiple federal agencies and programs that provide drinking water and sanitation services to tribes, and support tribal self-determination to ensure that tribes have greater involvement in the provision of safe drinking water on their homelands.[81]

When allocating funding and designing new programs, we need to make sure we are thinking about water affordability. Research estimates that the share of American households experiencing unaffordable water bills will spike from 12 percent to nearly 36 percent in the near future.[82] This is a considerable jump and means that one-third of Americans will face water bills they cannot afford to pay. Important next steps include better understanding the scope of affordability and advancing appropriate programs of customer assistance.[83] Some US cities have been experimenting with innovative methods of billing, including Philadelphia, Baltimore, and Portland, so that low-income residents can afford water service and do not suffer shutoff.[84] Creative investments in rural communities will protect the health and safety of rural dwellers, along with stimulating economic growth by improving quality of life and attracting jobs to the area.[85]

Strategic investment demands good data. Too few data have been collected systematically across the United States regarding water affordability, or even access to water.[86] We lack national requirements and data collection to report on disconnections, for example.[87] At present, there is no federal agency or research institution that collects comprehensive data on water affordability, access, or shutoffs in the United States.[88]

Strategic investment demands that we apply an equity lens to reveal the links among systemic racism, poverty, and water in the United States.[89] The lack of infrastructure for both rural and urban Black neighborhoods is a historical legacy that can be traced back to New Deal–era construction, stretching beyond more modern federal disinvestment.[90] This lack of adequate infrastructure is often a vestige of Jim Crow–era laws of discriminatory zoning and land-use regulation, which resulted in many rural African American communities lacking basic sewage systems and a trustworthy water supply.[91] Decades of racialized urban planning can be seen as a root cause of water inequalities whereby wealthier, whiter neighborhoods have historically attracted greater public and private investment.[92] An equity approach emphasizes fairness in water-related decision-making. This implies granting full access to information and the ability to participate in water-related decisions and outcomes by correcting imbalances in power, access, and distributive fairness.[93] For sure, our national reinvestment in water will demand creativity, experimentation, and collaboration. We are not short on good ideas; rather, what we often lack are political leadership and social mobilization.

How to Read This Book

It is our responsibility—and privilege—to better understand the depth and breadth of our water challenges. For far too long, water has been treated as a technical issue relegated to the engineers and scientists working for municipalities, or it has been isolated in academic journals and books not especially accessible to the public. But the issues surrounding water today are far too important to be confined to such narrow spaces and limited participants. This volume is intended to be a modern guide to water for readers, one that will equip them with the knowledge to see these water challenges as deeply social and political and not merely technical in nature.

To achieve this goal, we selected the best writing and thinking on critical water-related issues published in The Conversation over the past few years. These high-quality contributions encompass a range of quality and supply problems, including affordability, climate change, health and pollution risks, and the ocean environment. Readers will come to appreciate the interconnectedness of water with food, energy, and biodiversity. The chapters collected herein highlight the many dimensions of the water crisis in the United States today and also shine light on these issues playing out in other parts of the world. The writing is accessible and timely, capturing and distilling the latest science, technological achievements, and political and regulatory environment of water.

This volume reveals the multifaceted challenges ahead but also calls attention to potential solutions that provide opportunities for hope. New partnerships and collaborative efforts are under way in many parts of the world, revealing alternative governance practices and greater inclusivity in how we manage water. New scientific tools and citizen-science approaches are empowering local communities to collect their own data with which to manage their water resources better. There is much to be hopeful about—but more work needs to be done.

Ultimately, the chapters in this volume provide readers with a critical context for understanding the issues that humanity faces today with regard to water. Understanding the challenges before us is the first step to building a more sustainable and equitable water future for all. It is my sincerest hope that readers will find these readings informative and inspiring and will use them to promote change and justice in their communities.

Notes

1. Carla Campbell, Rachael Greenberg, Deepa Mankikar, and Ronald D. Ross, "A Case Study of Environmental Injustice: The Failure in Flint," *International Journal of Environmental Research and Public Health* 13, no. 10 (2016): 1–11; Benjamin J. Pauli, "The Flint Water Crisis," *WIREs Water* 7, no. 3 (2020): e1420.
2. Justin Worland, "America's Clean Water Crisis Goes Far beyond Flint. There's No Relief in Sight," *Time*, February 20, 2020, https://time.com /longform/clean-water-access-united-states/.
3. Kristi Pullen Fedinick, Steve Taylor, and Michele Roberts, *Watered Down Justice* (Washington, DC: Natural Resources Defense Council / Environmental Justice Health Alliance, 2020).
4. Fedinick, Taylor, and Roberts, *Watered Down Justice*.
5. *The Clean Water Act at 50: Promises Half Kept at the Half-Century Mark* (Washington, DC: Environmental Integrity Project, March 17, 2022),

https://environmentalintegrity.org/wp-content/uploads/2022/03/CWA
-report-3.23.22-FINAL.pdf.

6. American Society of Civil Engineers, "Report Card for America's Infrastructure," 2021, https://infrastructurereportcard.org/cat-item/drinking-water/.

7. Katie Meehan, Jason R. Jurjevich, Nicholas M. J. W. Chun, and Justin
Sherrill, "Geographies of Insecure Water Access and the Housing-Water
Nexus in US Cities," *Proceedings of the National Academy of Sciences of
the United States of America* 117, no. 46 (2020): 28700–707.

8. Zoë Roller, Stephen Gasteyer, Nora Nelson, WenHua Lai, and Marie Carmen
Shingne, *Closing the Water Access Gap in the United States: A National
Action Plan* (Washington, DC: Dig Deep / US Water Alliance, 2019).

9. Kevin Shafer and Radhika Fox, *An Equitable Water Future* (Washington,
DC: US Water Alliance, 2017).

10. *Lancet*, "The *Lancet* Commission on Pollution and Health," October 19,
2017, https://www.thelancet.com/commissions/pollution-and-health.

11. Chris Tyree and Dan Morrison, "Invisible: The Plastics inside Us," Orb
Media, 2017, https://orbmedia.org/the-invisibles; A. Dick Vethaak and
Juliette Legler, "Microplastics and Human Health," *Science* 371, no. 6530
(2021): 672–74.

12. Environmental Working Group, *Drinking Water and Children's Health*, July
27, 2017, https://www.ewg.org/research/drinking-water-and-childrens
-health.

13. James Conca, "Super Majority of Americans Worry about Clean Drinking
Water," *Forbes*, June 29, 2017, https://www.forbes.com/sites/jamesconca
/2017/06/29/super-majority-of-americans-worry-about-clean-drinking
-water/?sh=7cf161fd41e8.

14. Food and Water Watch, *America's Secret Water Crisis* (Washington, DC:
Food and Water Watch 2018), 1.

15. Shafer and Fox, *Equitable Water Future*.

16. Elizabeth A. Mack and Sarah Wrase, "Correction: A Burgeoning Crisis? A
Nationwide Assessment of the Geography of Water Affordability in the
United States," *PLOS One* 12, no. 4 (2017): e0176645.

17. Nina Lakhani, "Millions of Americans Can't Afford Water, as Bills Rise 80%
in a Decade," *Consumer Reports*, July 10, 2020, https://www.consumer
reports.org/personal-finance/millions-of-americans-cant-afford-water
-as-bills-rise-80-percent-in-a-decade-a8273700709/.

18. Newsha K. Ajami and Joseph W. Kane, *The Hidden Role of Water Infrastructure in Driving a COVID-19 Recovery*, Brookings Institution, 2020,
https://www.brookings.edu/blog/the-avenue/2020/10/20/the-hidden
-role-of-water-infrastructure-in-driving-a-covid-19-recovery/.

19. Richard J. Dawson, "Public Health Is Moot without Water Security," *Nature*
583, no. 7816 (2020): 360–61.

20. Brett Walton, "Water Shutoffs Are Suspended, but the Bills Will Still Be
Due," Circle of Blue, 2020, https://www.circleofblue.org/2020/world/water
-shutoffs-are-suspended-but-the-bills-will-still-be-due/?mc_cid
=7ff0999f1b&mc_eid=e4e8fd57ea; Lillian Holmes, Morgan Shimabuku,
Laura Feinstein, Greg Pierce, Peter H. Gleick, and Sarah Diringer, *Water*

and the COVID-19 Pandemic: Equity Dimensions of Utility Disconnections in the U.S. (Oakland, CA: Pacific Institute, 2020).

21. Brett Walton, "Congress Adds $638 Million in Water-Bill Debt Relief to Coronavirus Package," Circle of Blue, 2021, https://www.circleofblue.org/2020/world/congress-adds-638-million-in-water-bill-.

22. María Inés Zamudio, "Black Neighborhoods in Chicago Have Water Debt 10 Times Higher than White Areas," NPR WBEZ Chicago, January 18, 2022, https://www.wbez.org/stories/chicago-water-debt-10-times-higher-in-black-communities/129efc1a-70aa-4d3e-9ab3-46935343af0f?mc_cid=4a56c72617&mc_eid=e4e8fd57ea.

23. American Society of Civil Engineers and Value of Water Campaign, *The Economic Benefits of Investing in Water Infrastructure: How a Failure to Act Would Affect the US Economic Recovery*, 2020, https://www.infrastructureusa.org/the-economic-benefits-of-investing-in-water-infrastructure-2/.

24. Fedinick, Taylor, and Roberts, *Watered Down Justice*.

25. Roller, Gasteyer, Nelson, Lai, and Shingne, *Closing the Water Access Gap*.

26. Holmes, Shimabuku, Feinstein, Pierce, Gleick, and Diringer, *Water and the COVID-19 Pandemic*.

27. Meehan, Jurjevich, Chun, and Sherrill, "Geographies of Insecure Water Access."

28. James McBride and Anshu Siripurapu, "The State of U.S. Infrastructure," Council on Foreign Relations, November 8, 2021, https://www.cfr.org/backgrounder/state-us-infrastructure?gclid=CjwKCAiA6seQBhAfEiwAvPqu199x3JhVkJxUcRx1GhDQsbGfc3fFnDs-XXFn_tBz7dKnzvePOCMicBoCRQ8QAvD_BwE.

29. American Society of Civil Engineers, "Report Card for America's Infrastructure."

30. American Society of Civil Engineers, "Report Card for America's Infrastructure."

31. Juliet Christian-Smith, Peter Gleick, Heather Cooley, Lucy Allen, Amy Vanderwarker, and Katie Berry, *A Twenty-First Century US Water Policy* (Oakland, CA: Pacific Institute, 2012); Brett Walton, "Leaky Sewers Plunge Small North Carolina Towns into Financial Crisis," Circle of Blue, October 17, 2019, https://www.circleofblue.org/2019/world/leaky-sewers-plunge-small-north-carolina-towns-into-financial-crisis/?mc_cid=d88190db93&mc_eid=e4e8fd57ea.

32. Shafer and Fox, *Equitable Water Future*.

33. Andrea K. Gerlak, "Rural America Left Out of Trump's Water Infrastructure Plan," The Hill (blogs), June 13, 2017, http://thehill.com/blogs/pundits-blog/energy-environment/337636-rural-america-left-out-of-trumps-water-infrastructure.

34. WaterWorld, "Mind the Gap: Is WIFIA the Answer for US Water Difficulties?," November 1, 2013, https://www.waterworld.com/international/wastewater/article/16201809/mind-the-gap-is-wifia-the-answer-for-us-water-difficulties.

35. US Government Accountability Office, *Rural Water Infrastructure: Federal Agencies Provide Funding but Could Increase Coordination to Help Communities* (Washington, DC: GAO, 2015).

36. American Water Works Association, *Buried No Longer: Confronting America's Water Infrastructure Challenge* (Washington, DC: AWWA, 2017).

37. Xindi C. Hu, David Q. Andrews, Andrew B. Lindstrom, Thomas A. Bruton, Laurel A. Schaider, Philippe Grandjean, Rainer Lohmann, et al., "Detection of Poly- and Perfluoroalkyl Substances (PFASs) in U.S. Drinking Water Linked to Industrial Sites, Military Fire Training Areas, and Wastewater Treatment Plants," *Environmental Science & Technology Letters* 3, no. 10 (2016): 344–50.

38. Annie Sneed, "Forever Chemicals Are Widespread in U.S. Drinking Water," *Scientific American*, January 22, 2021, https://www.scientificamerican .com/article/forever-chemicals-are-widespread-in-u-s-drinking-water/.

39. Quoted in Brooks Hays, "Elevated PFAS Levels Found in Tap Water in Major U.S. Cities," UPI, January 22, 2020, https://www.upi.com/Science _News/2020/01/22/Elevated-PFAS-levels-found-in-tap-water-in-major -US-cities/5051579718169/.

40. Sneed, "Forever Chemicals Are Widespread in U.S. Drinking Water."

41. Hays, "Elevated PFAS Levels Found in Tap Water in Major U.S. Cities."

42. Scott Faber, "For 20-Plus Years, EPA Has Failed to Regulate 'Forever Chemicals,'" Environmental Working Group, January 9, 2020, https:// www.ewg.org/epa-pfas-timeline/.

43. Brady Dennis and Darryl Fears, "Biden Administration Moves to Curtail Toxic 'Forever Chemicals,'" *Washington Post*, October 18, 2021, https:// www.washingtonpost.com/climate-environment/2021/10/18/epa -regulate-forever-chemicals-pfas/.

44. Devin Dwyer, Stephanie Ebbs, and Jacqueline Yoo, "'Ticking Time Bomb': PFAS Chemicals in Drinking Water Alarm Scientists over Health Risks," ABC News, August 10, 2021, https://abcnews.go.com/US/ticking-time -bomb-pfas-chemicals-drinking-water-alarm/story?id=79300094#:~ :text=Last%20month%2C%20the%20city%20of,to%20overwhelm%20 groundwater%20filtration%20systems.

45. Jeanne Nienaber Clarke and Andrea K. Gerlak, "Environmental Racism in the Sunbelt: A Cross-Cultural Analysis," *Environmental Management* 22, no. 6 (1998): 857–67.

46. Andrea K. Gerlak and Jeanne Nienaber Clarke, "Tucson Opinion: With South-Side PFAS Water Contamination, History Repeats Itself," *Arizona Daily Star*, July 14, 2021, https://tucson.com/opinion/local/tucson -opinion-with-south-side-pfas-water-contamination-history-repeats -itself/article_d0aa9748-e40a-11eb-bc9a-f7abe1946fe1.amp.html.

47. D. Wuebbles, D. W. Fahey, and K. A. Hibbard, "How Will Climate Change Affect the United States in Decades to Come?," Eos, November 3, 2017, https://eos.org/features/how-will-climate-change-affect-the-united -states-in-decades-to-come.

48. AP News, "US Projections on Drought-Hit Colorado River Grow More Dire," September 22, 2021, https://apnews.com/article/business -environment-and-nature-arizona-lakes-colorado-river-8ae38d6247128 c61bbf1f917b075877d.

49. Lauren Sommer, "The Drought in the Western U.S. Is Getting Bad. Climate

Change Is Making It Worse," NPR, June 9, 2021, https://www.npr.org/2021/06/09/1003424717/the-drought-in-the-western-u-s-is-getting-bad-climate-change-is-making-it-worse.

50. H. Fountain, "In a First, U.S. Declares Shortage on Colorado River, Forcing Water Cuts," *New York Times*, August 16, 2021.

51. UNESCO World Water Assessment Programme, *The United Nations World Water Development Report, 2019: Leaving No One Behind* (Paris: UNESCO, 2019).

52. World Health Organization and United Nations Children's Fund, *Progress on Household Drinking Water, Sanitation and Hygiene, 2000–2017: Special Focus on Inequalities* (New York: WHO and UNICEF, 2019).

53. UNESCO World Water Assessment Programme, *United Nations World Water Development Report, 2019.*

54. World Meteorological Organization, *United in Science 2020: A Multi-organization High-Level Compilation of the Latest Climate Science Information* (Geneva: WMO, 2020).

55. UNESCO World Water Assessment Programme, *The United Nations World Water Development Report, 2020: Water and Climate Change* (Paris: UNESCO, 2020).

56. Fiona Harvey, "Poor Water Infrastructure Puts World at Greater Risk from Coronavirus," *Guardian*, March 21, 2020, https://www.theguardian.com/environment/2020/mar/22/water-saving-an-important-but-ignored-weapon-in-solving-climate-crisis-says-un.

57. United Nations MDG Monitor, *MDG 7: Ensure Environmental Sustainability*, June 6, 2017, http://www.mdgmonitor.org/mdg-7-ensure-environmental-sustainability/.

58. Andrea K. Gerlak, Madeline Baer, and Paula Lopes, "Taking Stock of the Human Right to Water," *International Journal of Water Governance* 6 (2018): 108–34.

59. "Goal 6: Ensure Availability and Sustainable Management of Water and Sanitation for All," Sustainable Development Goals, https://unstats.un.org/sdgs/report/2017/goal-06/.

60. United Nations, "As with So Many Other Human Rights, Indigenous Peoples Suffer Disproportionate Violations of Right to Safe Water, Sanitation, Permanent Forum Told," May 24, 2011, https://www.un.org/press/en/2011/hr5061.doc.htm.

61. Jamison Ervin, "Indigenous Communities: Advocates in Providing Clean Water for All," UNDP (blog), March 25, 2019, https://www.undp.org/blog/indigenous-communities-advocates-providing-clean-water-all.

62. United Nations Children's Fund, *For Every Child, Every Right: The Convention on the Rights of the Child at a Crossroads* (New York: UNICEF, 2019).

63. Victoria A. Beard and Diana Mitlin, "Water Access in Global South Cities: The Challenges of Intermittency and Affordability," *World Development* 147 (2021): 105625.

64. Keith Schneider, "Innovation in Financing Brightens WASH Galaxy," Circle of Blue, April 26, 2021, https://www.circleofblue.org/2021/world/innovation-in-financing-brightens-wash-galaxy/.

65. Nadya Ivanova, "Singapore Taps Waste Water and Tops Water Innovation," Circle of Blue, June 23, 2009, https://www.circleofblue.org/2009/world /singapore-tops-water-innovation/; Michael Taylor, "Thirsty Singapore Taps into Innovation to Secure Its Water Future," Reuters, June 7, 2019, https:// www.reuters.com/article/us-water-singapore-innovation/thirsty-singapore -taps-into-innovation-to-secure-its-water-future-idUSKCN1T8OVU.

66. Mark T. Buntaine, Bing Zhang, and Patrick Hunnicutt, "Citizen Monitoring of Waterways Decreases Pollution in China by Supporting Government Action and Oversight," *Proceedings of the National Academy of Sciences* 118, no. 29 (2021): e2015175118.

67. Pauli, "Flint Water Crisis."

68. Jennifer S. Carrera and Kent Key, "Troubling Heroes: Reframing the Environmental Justice Contributions of the Flint Water Crisis," *WIREs Water* 8 (2021): e1524.

69. Christian-Smith, Gleick, Cooley, Allen, Vanderwarker, and Berry, *Twenty-First Century US Water Policy*; Shafer and Fox, *Equitable Water Future*; Food and Water Watch, *America's Secret Water Crisis* (Washington, DC: Food and Water Watch, 2018).

70. David L. Feldman, *The Governance of Water Innovations: To Quench a Thirst* (Cheltenham, UK: Edward Elgar Publishers, 2022).

71. Brett Walton, "Water Utilities Call on Big Data to Guide Pipe Replacements," Circle of Blue, January 24, 2019, https://www.circleofblue.org /2019/world/water-utilities-call-on-big-data-to-guide-pipe-replacements /; Keith Schneider, "Three Thirsty Texas Cities Are Global Leaders in Water Innovation," Circle of Blue, August 17, 2020, https://www.circle ofblue.org/2020/world/three-thirsty-texas-cities-are-global-leaders -in-water-innovation/.

72. Leland Jackson, "We Brewed Beer from Recycled Wastewater—and It Tasted Great," The Conversation, November 16, 2020, https:// theconversation.com/we-brewed-beer-from-recycled-wastewater-and -it-tasted-great-148386; Alex Robinson, "Turning Toilet Water into Beer," Modern Farmer, August 24, 2020, https://modernfarmer.com/2020/08 /turning-toilet-water-into-beer/.

73. Schneider, "Three Thirsty Texas Cities Are Global Leaders in Water Innovation."

74. C. Staddon, L. De Vito, A. A. Zuniga-Teran, Y. Schoeman, A. Hart, and G. Booth, "Contributions of Green Infrastructure to Urban Resilience." Report elaborated by the consortium of the University of the West of England; University of Arizona; Monash South Africa University, Arcadis; and University of Science and Technology, Beijing, China, 2017.

75. Andrea K. Gerlak, Alison Elder, Mitch Pavao-Zuckerman, Adriana Zuniga-Teran, and Andrew R. Sanderford, "Agency and Governance in Green Infrastructure Policy Adoption and Change," *Journal of Environmental Policy & Planning* 23 no. 5 (2021): 599–615.

76. Kristi Pullen Fedinick, Mae Wu, Mekela Panditharatne, and Erik D. Olson, *Threats on Tap: Widespread Violations Highlight Need for Investment in Water Infrastructure and Protections* (Washington, DC: Natural Resources Defense Council, 2017).

77. Conca, "Super Majority of Americans Worry about Clean Drinking Water."
78. American Society of Civil Engineers and Value of Water Campaign, *Economic Benefits of Investing in Water Infrastructure*.
79. A. Bhatia and Q. Bui, "The Infrastructure Plan: What's In and What's Out," *New York Times*, August 10, 2021; White House, "Fact Sheet: The Bipartisan Infrastructure Deal," November 6, 2021, https://www.whitehouse.gov/briefing-room/statements-releases/2021/11/06/fact-sheet-the-bipartisan-infrastructure-deal/.
80. Gillian Flaccus, Felicia Fonseca, and Becky Bohrer, "US Tribes See Hope for Clean Water in Infrastructure Bill," *U.S. News and World Report*, December 23, 2021, https://www.usnews.com/news/politics/articles/2021-12-23/infrastructure-bill-to-aid-us-tribes-with-water-plumbing.
81. Water and Tribes Initiative, *Universal Access to Clean Water for Tribes: Recommendations for Operational, Administrative, Policy, and Regulatory Reform*, November 2021, https://tribalcleanwater.org/wp-content/uploads/2021/11/Full-Report-11.21-FINAL.pdf.
82. Mack and Wrase, "Correction."
83. Christian-Smith, Gleick, Cooley, Allen, Vanderwarker, and Berry, *Twenty-First Century US Water Policy*.
84. Brett Walton, "Baltimore Council Approves Income-Based Water Bills," Circle of Blue, November 21, 2019, https://www.circleofblue.org/2019/world/baltimore-council-approves-income-based-water-bills/.
85. Gerlak, "Rural America Left Out of Trump's Water Infrastructure Plan."
86. Roller, Gasteyer, Nelson, Lai, and Shingne, *Closing the Water Access Gap in the United States*.
87. Holmes, Shimabuku, Feinstein, Pierce, Gleick, and Diringer, *Water and the COVID-19 Pandemic*; Food and Water Watch, *America's Secret Water Crisis*; Brett Walton, "Lack of Utility Data Obscures Customer Water Debt Problems," Circle of Blue, https://www.circleofblue.org/2020/world/lack-of-utility-data-obscures-customer-water-debt-problems/?mc_cid=c74ca0cb7b&mc_eid=e4e8fd57ea.
88. Fedinick, Taylor, and Roberts, *Watered Down Justice*.
89. Andrea K. Gerlak, Elena Louder, and Helen Ingram, "Viewpoint: An Intersectional Approach to Water Equity in the US," *Water Alternatives* 15, no. 1 (2022): 1–12.
90. Fedinick, Taylor, and Roberts, *Watered Down Justice*.
91. Christian-Smith, Gleick, Cooley, Allen, Vanderwarker, and Berry, *Twenty-First Century US Water Policy*.
92. Roller, Gasteyer, Nelson, Lai, and Shingne, *Closing the Water Access Gap in the United States*; Lauren Contorno, Mariana Sarango, and Sharon L. Harlan, *Environmental Justice and Sustainable Urban Water Systems: Community Voices from Selected Cities in the United States* (Boston: Northeastern University, 2018), https://erams.com/UWIN/wp-content/uploads/2018/10/UWIN_SEEJ_Report_October-2018.pdf.
93. Gerlak, Louder, and Ingram, "Viewpoint."

The Conversation on Water

Part I

Health and the Need for Clean Water

Water is essential to all life, and clean water is essential to good health. The Flint water crisis from the mid-to-late 2010s was a tragic wake-up call for many Americans who perhaps assumed that as long as their water flowed from a tap, it was safe to drink. But lead pipes, which contaminated the drinking water of Flint residents, are only one of a number of concerns for water quality linked to a legacy of insufficient funding for maintaining infrastructure, negligence in countering climate change, and failure to mitigate other emerging threats.

Today, communities across the United States face water-quality challenges associated with wildfires, harmful algal blooms, and emerging contaminants, like polyfluoroalkyl substances, or PFAS. There are serious public health concerns associated with drinking contaminated water. Lead poisoning increases the risk of kidney disease, stroke, and hypertension, and researchers have connected lead poisoning with incarceration for violent crimes. Long-term effects of exposure to the carcinogen benzene after wildfires can decrease a person's white blood cells, which protect the body from infectious disease. Algal blooms cause foul-smelling and -tasting water, and they release trace metals that may cause health problems. As PFAS concentrate in human organs, tissues, and cells, they can lead to testicular and kidney cancer, liver damage, decreased fertility, and thyroid disease. Exposure to PFAS even reduces the effectiveness of vaccines. Children are especially vulnerable. In writing about the Biden administration's plan to remove lead pipes from US homes, Filippelli explains, in his chapter of part I, that lead poisoning disproportionately affects communities of color, with Black, Hispanic, and Mexican American children bearing the highest elevated levels of lead in their blood.

Increasingly, Americans are experiencing heightened anxiety and stress associated with these water problems. The growing and understandable public distrust of government over water quality, writes Rosinger, leads to aggravated health issues, especially in vulnerable communities, and erodes faith in government overall. People wary of tap water look to alternative sources such as bottled water and sugary drinks, which of course have other public health implications. In some cases, households are not sufficiently warned of the dangers of contaminated water and the actions they can take to reduce its risks, as Whelton describes in his chapter on how western wildfires are associated with drinking water contamination. Sometimes regulations are not being followed properly. In other cases, appropriate regulations fail even to exist, despite significant public health research.

The chapters of part I explain the most pressing issues—and some solutions—for ensuring water quality and stemming public health concerns over contaminated drinking water. They reflect the great advances in public health research toward better understanding the scope and severity of water contamination risks for humans. They convey real hope in the advocacy work and public education campaigns emerging around these issues. This can be seen in the efforts led by scientists in calling for a comprehensive, effective plan for managing some of these contaminants, like PFAS, as outlined by Kwiatkowski. And hope can also be seen in partnerships formed to harness smart technologies that could address threats to the Great Lakes' water quality, according to Filippelli and Ortiz.

By examining the issues set forth in part I, we can gain a more comprehensive picture of what clean water in contemporary times really requires, and we can build on some truly collaborative and innovative approaches under way today in parts of the United States.

Chapter 1

Nearly 60 Million Americans Don't Drink Their Tap Water, Research Suggests—Here's Why That's a Public Health Problem

ASHER ROSINGER

IMAGINE SEEING A NEWS REPORT about lead contamination in drinking water in a community that looks like yours. It might make you think twice about drinking your tap water or serving it to your kids—especially if you also have experienced tap water problems in the past.

In a 2021 study, my colleagues Anisha Patel and Francesca Weaks and I estimated that approximately 61.4 million people in the United States did not drink their tap water as of 2017–2018.

In Jackson, Mississippi, residents pick up bottled water at a city distribution center on February 18, 2021. Much of the city was without safe drinking water because of problems at its water treatment plant.
AP Photo/Rogelio V. Solis

Our research, published in the journal *Public Health Nutrition*, found that this number has grown sharply in the past several years.[1]

Other research has shown that about 2 million Americans don't have access to clean water. Taking that into account, our findings suggest that about 59 million people have tap water access from either their municipality or private wells or cisterns, but they don't drink it. While some may have contaminated water, others may be avoiding water that's actually safe.

Water insecurity is an underrecognized but growing problem in the United States. Tap water distrust is part of the problem. And it's critical to understand what drives it, because people who don't trust their tap water shift to more expensive and often less healthy options, like bottled water or sugary drinks.

I'm a human biologist and have studied water and health for the past decade in places as diverse as Lowland Bolivia and northern Kenya. Now I run the Water, Health, and Nutrition Laboratory at Pennsylvania State University. To understand water issues, I talk to people and use large datasets to see whether a problem is unique or widespread, and stable or growing.

An Epidemic of Distrust

According to our research, there's a growing epidemic of tap water distrust and disuse in the United States. In a 2020 study, anthropologist Sera Young and I found that tap water avoidance was declining before the Flint water crisis that began in 2014. In 2015–2016, however, it started to increase again for children.[2]

Our study from 2021 found that, in 2017–2018, the number of Americans who didn't drink tap water increased at an alarmingly high rate, particularly for Black and Hispanic adults and children. Since 2013–2014—just before the Flint water crisis began—the prevalence of adults who do not drink their tap water has increased by 40 percent. Among children, not consuming tap water has risen by 63 percent.

To calculate this change, we used data from the National Health and Nutrition Examination Survey, a national survey that releases data in two-year cycles. Sampling weights that use demographic characteristics ensure that the people being sampled are representative of the broader US population.

Racial Disparities in Tap Water Consumption

Communities of color have long experienced environmental injustice across the United States. Black, Hispanic, and Native American residents are more likely to live in environmentally disadvantaged neighborhoods with exposure to water that violates quality standards.

Our findings reflect these experiences. We calculated that Black and Hispanic children and adults are two to three times more likely to report not drinking their tap water than are members of white households. In 2017–2018, roughly 3 out of 10 Black adults and children and nearly 4 of 10 Hispanic adults and children didn't drink their tap water. Approximately 2 of 10 Asian Americans didn't drink from their tap, while only 1 of 10 white Americans didn't drink their tap water.

When children don't drink any water on a given day, research shows that they consume twice as many calories from sugary drinks as do children who drink water.[3] Higher consumption of sugary drinks increases the risk of cavities, obesity, and cardiometabolic diseases. Drinking tap water provides fluoride, which lowers the risk of cavities. Relying on water alternatives is also much more expensive than drinking tap water.

What Erodes Trust

News reports—particularly of high-visibility events like advisories to boil water—lead people to distrust their tap water even after the problem is fixed. For example, a 2019 study showed that water quality violations across the United States between 2006 and 2015 led to increases in bottled water purchases in affected counties as a way to avoid tap water, and purchase rates remained elevated after the violation was resolved.[4]

The Flint water crisis drew national attention to water insecurity, even though state and federal regulators were slow to respond to residents' complaints there. Soon afterward, lead contamination was found in the water supply of Newark,

New Jersey; the city is currently replacing all lead service lines under a legal settlement. Elsewhere, media outlets and advocacy groups have reported finding tap water samples contaminated with industrial chemicals, lead, arsenic, and other contaminants.

Many other factors can cause people to distrust their water supply, including smell, taste, and appearance, as well as lower income levels. Location is also an issue: older US cities with aging infrastructure are more prone to water shutoffs and water quality problems.[5]

It's important not to blame people for distrusting what comes out of their tap, because those fears are rooted in history. In my view, addressing water insecurity requires a two-part strategy: ensuring that everyone has access to clean water and increasing trust so that people who have safe water will use it.

Building Confidence

As part of the 2021 infrastructure plan, President Joe Biden appropriated US$55 billion to improve water delivery systems, replace lead pipelines, and tackle other contaminants. The plan also sought to improve small water systems and underserved communities.

These are critical steps to rebuilding trust. Yet, in my view, the Environmental Protection Agency should also provide better public education about water quality testing and targeted interventions for vulnerable populations, such as children and underserved communities. Initiatives to simplify and improve water quality reports can help people understand what's in their water and what they can do if they think something is wrong with it.

Who is delivering those messages is important. In areas like Flint, where former government officials have been indicted on charges of negligence and perjury in connection with the water crisis, the government's word alone won't rebuild trust. Instead, community members can fill this critical role.

Another priority is the 13–15 percent of Americans who rely on private well water, which is not regulated under the Safe Drinking Water Act. These households are responsible for their own water quality testing. Public funding would help them test it regularly and address any problems.

Public distrust of tap water in the United States reflects decades of policies that have reduced access to reliable, safe drinking water in communities of color. Fixing water lines is important but so too is giving people confidence to turn on the tap.

Notes

1. Asher Y. Rosinger, Anisha I. Patel, and Francesca Weaks, "Examining Recent Trends in the Racial Disparity Gap in Tap Water Consumption: NHANES 2011–2018," *Public Health Nutrition* 25, no. 2 (2022): 207–13, https://doi.org/10.1017/s1368980021002603.
2. Asher Y. Rosinger and Sera L. Young, "In-Home Tap Water Consumption Trends Changed among U.S. Children, but Not Adults, between 2007 and 2016," *Water Resources Research* 56, no. 7 (2020): e2020WR027657, https://doi.org/10.1029/2020wr027657.
3. Asher Y. Rosinger, Hilary Bethancourt, and Lori A. Francis, "Association of Caloric Intake from Sugar-Sweetened Beverages with Water Intake among US Children and Young Adults in the 2011–2016 National Health and Nutrition Examination Survey," *JAMA Pediatrics* 173, no. 6 (2019): 602–604, https://doi.org/10.1001/jamapediatrics.2019.0693.
4. Maura Allaire, Taylor Mackay, Shuyan Zheng, and Upmanu Lall, "Detecting Community Response to Water Quality Violations Using Bottled Water Sales," *Proceedings of the National Academy of Sciences* 116, no. 42 (2019): 20917–22, https://doi.org/10.1073/pnas.1905385116.
5. Marian Swain, Emmett McKinney, and Lawrence Susskind, "Water Shutoffs in Older American Cities: Causes, Extent, and Remedies," *Journal of Planning Education and Research* (2020): https://doi.org/10.1177/0739456x20904431.

Chapter 2

The Importance of Replacing Lead Water Pipes from Coast to Coast

GABRIEL FILIPPELLI

THE BIDEN ADMINISTRATION in late 2021 released a plan to accelerate the removal of lead water pipes and lead paint from US homes.[1] As a geochemist and environmental health researcher who has studied the heartbreaking impacts of lead poisoning in children for decades, I am happy to see high-level attention paid to this silent killer, which disproportionately affects poor communities of color.

Childhood lead poisoning has declined significantly in the United States over the past 50 years. That's largely due to the elimination of leaded gasoline in the 1980s and the banning of most lead-based paints.

But the US Environmental Protection Agency estimates that up to 10 million households and 400,000 schools and childcare centers have service lines or other fixtures that contain lead. These pipes are ticking time bombs that can leach toxic lead into drinking water if they corrode. As long as they remain in service, children and families are vulnerable.

The same is true of lead paint, which is still present in many homes built before consumer use of lead paint was banned in 1978. Because it tastes sweet, children sometimes chew on paint chips or painted wood.

The Biden administration called for spending US$15 billion from the 2021 infrastructure bill to replace lead service lines, faucets, and fixtures over five years. I see this as a priority, since Black children and children living in poverty have average levels of blood lead that are 13 percent higher than the national average.

Lead Poisoning Does Permanent Damage

Lead poisoning is a major public health problem because lead has permanent impacts on the brain, particularly in children.

Lead poisoning disproportionately affects communities of color

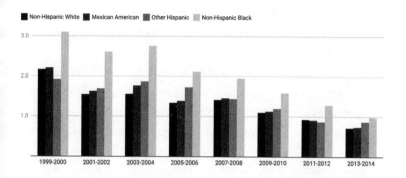

Young brains are still actively forming the amazing network of neurons that compose their hardware. Neurons are designed to use calcium, the most abundant mineral in the human body, as a transmitter to pass signals rapidly. Lead molecules look a lot like calcium molecules, so if they are present in a child's body, they can penetrate the brain, impair neuron development, and cause permanent neural damage.

Children with lead poisoning have lower IQs, poor memory recall, high rates of attention deficit disorder, and low impulse control. They tend to perform poorly at school, which reduces their earning potential as adults.[2] They also face increased risk of kidney disease, stroke, and hypertension as they age. Research has found strong connections between lead poisoning and incarceration for violent crimes.[3]

Researchers estimate that about 500,000 US children today have elevated blood lead levels.[4] Health experts widely agree that there is no known "safe" blood lead concentration.

Where Are the Lead Pipes?

The Biden administration's plan calls for replacing 100 percent of lead service lines across the nation. Step 1 is finding the pipes.

Most US cities have countless miles of lead service lines buried beneath streets and sidewalks and feeding into people's homes. Utilities don't know where many of these aging lines are and don't have enough data to map them. Replacing them will require significant analysis, modeling, data, and some guesswork. Old service lines have caused lead poisoning outbreaks in such places as Washington, DC; Flint, Michigan; and Newark, New Jersey. The chemistry is a bit different in each case.

Lead service lines typically develop a protective "plaque" of minerals on their inside walls after a short time, which effectively separates the toxic lead pipe from the water flowing through it. This coating, which is called scale, remains stable if the chemistry of the water coursing through it doesn't change. But if that chemistry is altered, disaster can ensue.

In 2002 Washington, DC, shifted from chlorine to chloramine for treating its water supply. Chloramine is a more modern disinfectant that does not form dangerous reactive chlorinated by-products as chlorine can. This rapidly corroded the protective plaque lining the city's pipes, flushing highly absorbable lead into homes. Tens of thousands of children were exposed over two years before the problem was adequately identified and fixed.

In Flint, state-appointed managers decided to save money during a fiscal crisis in 2014 by switching from Detroit water to water from the Flint River. But regulators did not require enough chemical analysis to determine what additives should be used to maintain the pipe plaque. And they skipped the typical step of adding phosphate, which binds chemically with lead and prevents it from leaching out of pipes, in order to save about $100 per day.[5]

Corrosion chemistry is well controlled in many US cities, but it is not a perfect science. And utilities don't always have detection systems that adequately alert water suppliers to dangers at the tap. That's why removing lead pipes is the only sure way to avoid the threat of more water crises.

Cities Will Need to Innovate

While US$15 billion is a big investment, experts agree that it's not enough to replace all lead pipes nationwide. For example, the estimated cost of replacing all of Flint's lead service lines is about US$50 million—and there are thousands of US cities to fix.

My own city, Indianapolis, has a population of around 850,000—about 10 times larger than Flint's—and officials have only a rough idea of where to find the lead service lines. There are ways to statistically model the likelihood that a given portion of the water system has lead service lines, using information such as water main sizes, locations, and construction dates, but they are imperfect.[6]

Cities will need to get creative to make whatever funds they get go as far as possible. As one example, I am working with colleagues to develop a citizen-science project that will provide thousands of tests for lead at taps around Indianapolis. This effort, a partnership with the University of Notre Dame funded by the US Department of Housing and Urban Development, should augment modeling with real data on levels of lead in homes and so increase public awareness of this issue.

In spite of these challenges, I believe urgency over this issue is long overdue. Every lead pipe that's replaced will pay off in higher lifetime earnings and lower rates of illness for families that gain access to safer tap water.

Notes

1. White House, "Fact Sheet: The Biden-Harris Lead Pipe and Paint Action Plan," December 16, 2021, https://www.whitehouse.gov /briefing-room/statements-releases/2021/12/16/fact-sheet-the -biden-harris-lead-pipe-and-paint-action-plan/.
2. Centers for Disease Control and Prevention, "Health Effects of Lead Exposure," last reviewed March 9, 2022, https://www.cdc.gov/nceh /lead/prevention/health-effects.htm.
3. Jennifer L. Doleac, "New Evidence That Lead Exposure Increases Crime," Brookings Institution, June 2, 2017, https://www.brookings .edu/blog/up-front/2017/06/01/new-evidence-that-lead-exposure -increases-crime/.
4. Laura Mayans, "Lead Poisoning in Children," *American Family Physician* 100, no. 1 (2019): 24–30, https://www.aafp.org/afp/2019/0701 /p24.html.
5. Jayde Lovell, "Q&A: What Really Happened to the Water in Flint, Michigan?," *Scientific American*, March 2, 2016, https://www .scientificamerican.com/article/q-a-what-really-happened-to -the-water-in-flint-michigan/.
6. Jake Abernethy and Eric Schwartz, "Statistical Modeling in Support of Lead Service Line Identification Inventory and Replacement," Water Online, January 13, 2021, https://www.wateronline.com/doc /statistical-modeling-in-support-of-lead-service-line-identification -inventory-and-replacement-0001.

Chapter 3

Wildfires Are Contaminating Drinking Water Systems, and It's More Widespread Than People Realize

ANDREW J. WHELTON

MORE THAN 58,000 WILDFIRES scorched the United States in 2020, and wildfire impacts are expected to become more intense and frequent. What many people don't realize is that these wildfires can do lasting damage beyond the reach of the flames—they can contaminate entire drinking water systems with carcinogens that last for months after the blaze. That water flows to homes, contaminating the plumbing, too.

Between 2017 and 2021, wildfires contaminated drinking water distribution networks and building plumbing for more than 240,000 people. Small water systems that serve housing developments, mobile home parks, businesses, and small towns have been particularly hard-hit. Most didn't realize their water was unsafe until weeks to months after the fire. The problem starts when wildfire smoke gets into the system or when plastic in water systems heats up. Heating can cause plastics to release harmful chemicals, like benzene, which can contaminate drinking water and permeate the system.

As an environmental engineer, I and my colleagues work with communities recovering from wildfires and other natural disasters. In 2020, at least seven water systems were found to be contaminated, suggesting that drinking water contamination may be a more widespread problem than people realize. Our 2021 study identifies critical issues that households and businesses should consider after a wildfire.[1] Failing to address them can harm people's health: mental, physical, and financial.

Wildfires Make Drinking Water Unsafe

When wildfires damage water distribution pipes, wells, and the plumbing in homes and other buildings, they can create immediate health risks. A building's plumbing can become contaminated by smoke getting sucked into water systems, by heat damaging plastic pipes,[2] or by contamination penetrating into the plumbing and leaching out slowly over time. Since 2017, multiple fires have rendered drinking water systems unsafe, including the Echo Mountain, Lionshead, and Almeda Fires in Oregon, and the CZU Lightning Complex, Camp, and Tubbs Fires in California.

Thousands of Private Wells Have Been Affected Too

Being exposed to contaminated water can cause immediate harm, such as headaches, nausea, dizziness, and vomiting.

Short-term exposure to 26 parts per billion or more of benzene, a carcinogen, may cause a decrease in white blood cells that protect the body from infectious disease. Multiple fires have caused drinking water to exceed this level. A variety of other chemicals can also exceed the exposure limits for safe drinking water in the absence of benzene.[3]

Households Are Not Being Adequately Warned

In a survey of 233 households affected by water contamination, we found that people reported high levels of anxiety and stress linked to the water problems. Nearly half had installed in-home water treatment because of uncertainty about the water's quality. Eighty-five percent had looked for other water sources, such as bottled water.

In some cases, we found that advice from government agencies placed households at greater risk of harm. It has sometimes left people exposed to chemicals, caused them to spend money needlessly, or given them a false sense of security. Certified in-home water treatment devices, for example, are tested only to bring down 15 parts per billion of benzene to less than 5 parts per billion, the federal standard. These devices are not tested to treat hazardous waste–scale contaminated water that's been found after wildfires.

Following the 2020 CZU Lightning Complex Fire near Santa Cruz, California, a local health department warned private well owners not to use their water, to test it, and to use a nearby damaged water system instead, and the state did not warn 17,000 people against bathing in the contaminated water. It was only after test results proved the water had been unsafe all along that the system owner and state advised against bathing in it.

In Oregon, some damaged systems encouraged people to boil their drinking water, later finding out that the water had benzene in it.

After the 2018 Camp Fire that devastated Paradise, California, the local health department correctly warned the entire county not to use or try to treat the drinking water, which had contamination above the hazardous waste limit of the US Environmental Protection Agency. But one water system and the state encouraged 13,000 people to try to treat it themselves.[4]

In all of these cases, the US Environmental Protection Agency chose not to compel water utilities to explicitly notify customers about the water contamination and its risk.

Communities have received other bad information:

- Commercial labs and government officials recommended flushing faucets for 5 to 15 minutes before collecting a water sample, thereby dumping out the contaminated plumbing water meant for testing.
- Homeowners were led to believe a single cold-water sample at the kitchen sink would determine if the hot water system and property service line were contaminated. It cannot.
- People were led to believe that benzene water testing would determine if any other chemicals were present above safe limits. This is not possible.

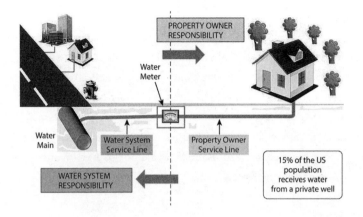

PROPERTY OWNER
RESPONSIBILITY

Water
Meter

Water
Main

Water System
Service Line

Property Owner
Service Line

WATER SYSTEM
RESPONSIBILITY

15% of the US
population
receives water
from a private well

What to Look For after a Nearby Fire

Signs of potential contamination after a nearby wildfire could
be a loss of water pressure, discolored water, heat damage to
water systems inside and outside buildings, and broken and
leaking pipes, valves, and hydrants.

Drinking water should be assumed to be chemically unsafe until proven otherwise.

Once a system is contaminated, cleanup can take months.
The water system will have to be flushed and tested regularly
to track down contamination.[5] Health departments should
also issue guidance on how to test private wells and plumbing.

When testing plumbing, include the property service
line as well as the hot- and cold-water lines. Before a water
sample is collected, the water must sit long enough in the
plumbing so that contamination can be found—72 hours was
the Tubbs Fire and Camp Fire standard. Tests should look for
more than benzene alone.

Who Can Help?

Many of the critical public health risks identified in our study can be addressed by public health departments with financial support from state and local agencies. Public health departments often have experience responding to water problems, such as legionella outbreaks, and can provide technical advice about chemical exposures, building plumbing, and private wells of drinking water.

Notes

1. Tolulope O. Odimayomi, Caitlin R. Proctor, Qi Erica Wang, Arman Sabbaghi, Kimberly S. Peterson, David J. Yu, Juneseok Lee, et al., "Water Safety Attitudes, Risk Perception, Experiences, and Education for Households Impacted by the 2018 Camp Fire, California," *Natural Hazards* 108, no. 1 (2021): 947–75, https://doi.org/10.1007/s11069-021-04714-9.

2. Kristofer P. Isaacson, Caitlin R. Proctor, Q. Erica Wang, Ethan Y. Edwards, Yoorae Noh, Amisha D. Shah, and Andrew J. Whelton, "Drinking Water Contamination from the Thermal Degradation of Plastics: Implications for Wildfire and Structure Fire Response," *Environmental Science: Water Research & Technology* 7, no. 2 (2021): 274–84, https://doi.org/10.1039/d0ew00836b.

3. Caitlin R. Proctor, Juneseok Lee, David Yu, Amisha D. Shah, and Andrew J. Whelton, "Wildfire Caused Widespread Drinking Water Distribution Network Contamination," *AWWA Water Science* 2, no. 4 (2020): https://doi.org/10.1002/aws2.1183.

4. *Information to Water Customers regarding Water Quality in Buildings Located in Areas Damaged by Wildfire*, California Water Boards, August 9, 2019, https://www.waterboards.ca.gov/drinking_water/programs/documents/benzenecustomeradvisoryfinal.pdf.

5. *Job Aid for Disaster Recovery Reform Act, Section 1205 Additional Activities for Wildfire and Wind Implementation under Hazard Mitigation Assistance Programs*, Federal Emergency Management Agency, December 3, 2019, https://www.fema.gov/sites/default/files/2020-07/fema_DRRA-1205-implementation-job-aid.pdf.

Chapter 4

Climate Change Threatens Drinking Water Quality across the Great Lakes

GABRIEL FILIPPELLI and JOSEPH D. ORTIZ

"DO NOT DRINK / DO NOT BOIL" is not what anyone wants to hear about their city's tap water. But the combined effects of climate change and degraded water quality could make such warnings more frequent across the Great Lakes region.

A preview occurred on July 31, 2014, when a nasty green slime—properly known as a harmful algal bloom, or HAB—developed in the western basin of Lake Erie. Before long it had overwhelmed the Toledo Water Intake Crib, which provides drinking water to nearly 500,000 people in and around the city.

Tests revealed that the algae were producing microcystin, a sometimes deadly liver toxin and suspected carcinogen. Unlike some other toxins, microcystin can't be rendered harmless by boiling. So the city issued a "Do Not Drink / Do Not Boil" order that set off a three-day crisis.[1] Local stores soon ran out of bottled water. Ohio's governor declared a state of emergency, and the National Guard was called in to provide safe drinking water until the system could be flushed and treatment facilities brought back on-line.

The culprit was a combination of high nutrient pollution—nitrogen and phosphorus, which stimulate the growth of algae—from sewage, agriculture, and suburban runoff, and high water temperatures linked to climate change. This event showed that even in regions with resources as vast as the Great Lakes, water supplies are vulnerable to these kinds of man-made threats.

As midwesterners working in the fields of urban environmental health and climate and environmental science, we believe more crises like Toledo's could lie ahead if the region doesn't address looming threats to drinking water quality.

Vast and Abused

The Great Lakes together hold 20 percent of the world's surface freshwater—more than enough to provide drinking water to over 48 million people from Duluth to Chicago, Detroit, Cleveland, and Toronto. But human impacts have severely harmed this precious and vital resource.

In 1970, after a century of urbanization and industrialization around the Great Lakes, water quality was severely degraded. Factories were allowed to dump waste into waterways rather than treating it. Inadequate sewer systems often sent raw sewage into rivers and lakes, fouling the water and causing algal blooms.

Problems like these helped spur two major steps in 1972: passage of the US Clean Water Act and the adoption of the Great Lakes Water Quality Agreement between the United States and Canada. Since then, many industries have been cleaned up or shut down. Sewer systems are being redesigned, albeit slowly and at great cost.

The resulting cuts in nutrient and wastewater pollution have brought a quick decline in HABs—especially in Lake Erie, the Great Lake with the most densely populated shoreline. But new problems have emerged, due partly to shortcomings in that law and agreement, combined with the growing effects of climate change.

Warmer and Wetter

Climate change is profoundly altering many factors that affect life in the Great Lakes region. The most immediate impacts of recent climate change have been on precipitation, lake levels, and water temperatures.

Annual precipitation in the region has increased by about 5 inches over the past century.[2] It's not just the amount of precipitation but how fast it falls that is important for water quality. Extreme precipitation events increased by 40 percent in the Midwest from 1958 to 2016.[3] As a consequence of both of these factors, the Great Lakes are increasingly experiencing extremely high and rapidly varying water levels.

Record high precipitation in 2019 caused flooding, property damage, and beachfront losses in a number of lakeside communities. Precipitation in 2020 was projected to be equally high if not higher. Some of this is due to natural variability, but certainly some is due to climate change.

Another clear impact of climate change is a general

warming of all five Great Lakes, particularly in the springtime. The temperature increase is modest and varies from year to year and place to place, but it is consistent overall with records of warming throughout the region.

More Polluted Runoff

Some of these climate-related changes have converged with more direct human impacts to influence water quality in the Great Lakes.

Cleanup measures adopted back in the 1970s imposed stringent limits on large point sources of nutrient pollution, like wastewater and factories. But smaller "nonpoint" sources, such as fertilizer and other nutrients washing off farm fields and suburban lawns, were addressed through weaker, voluntary controls. These have since become major pollution sources.

Since the mid-1990s, climate-driven increases in precipitation have carried growing quantities of nutrient runoff into Lake Erie. This rising load has triggered increasingly severe algal blooms, comparable in some ways to the events of the 1970s. Toledo's 2014 crisis was not an anomaly.

These blooms can make lake water smell and taste bad, and sometimes they make it dangerous to drink. They also have long-term impacts on the lakes' ecosystems. They deplete oxygen, killing fish and spurring chemical processes that prime the waters of Lake Erie for larger future blooms. Low-oxygen water is more corrosive and can damage water pipes, causing poor taste or foul odor, and it helps release trace metals that may also cause health problems.

To make matters worse, much of these environmental and climate policies were rolled back during the Trump administration.[4] So despite a half century of advances, in many ways water quality in the Great Lakes is back to where it was in 1970 but with the added influence of a rapidly changing climate.

Filtering Runoff

How can the region change course and build resilience in communities around the Great Lakes? Thanks to a number of studies, including an intensive modeling analysis of future climate change in Indiana, which serves as a proxy for most of the region, we have a pretty good picture of what the future could look like.

As one might guess, warming will continue. Summertime water temperatures are projected to rise by about another 5 degrees Fahrenheit by midcentury, even if nations significantly reduce their greenhouse gas emissions. This will cause further declines in water quality and negatively impact coastal ecosystems.

The analysis also projects an increase in extreme precipitation and runoff, particularly in the winter and spring. These shifts will likely bring still more nutrient runoff, sediment contaminants, and sewage overflows into coastal zones, even if surrounding states hold the actual quantities of these nutrients steady. More contaminants, coupled with higher temperatures, can trigger HABs that threaten water supplies.

But several success stories point to strategies for tackling these problems, at least at the local and regional levels.

A number of large infrastructure projects have focused on improving stormwater management and municipal sewer systems so that they can capture and process sewage and associated nutrients before they are transported to the Great Lakes. These initiatives help control flooding and increase the supply of "gray water"—or used water from bathroom sinks, washing machines, tubs, and showers—for uses such as landscaping.

Cities are coupling this gray infrastructure with green infrastructure projects, such as green roofs, infiltration gardens, and reclaimed wetlands. These systems can filter water to help remove excess nutrients. They also will slow runoff during extreme precipitation events, thus recharging natural reservoirs.

Additionally, municipal water managers are using smart technologies and improved remote sensing methods to create near-real-time warning systems for HABs that might help avert multiple crises. Groups like the Cleveland Water Alliance, an association of industry, government, and academic partners, are working to implement smart lake technologies in Lake Erie and other freshwater environments around the globe. Finally, states including Ohio and Indiana are moving to cut total nutrient inputs into the Great Lakes from all sources and to using advanced modeling to pinpoint those sources.[5]

Together, these developments could reduce the size of HABs and perhaps even reach the roughly 50 percent reduction in nutrient runoff that government studies suggest is needed to bring runoffs back to their minimum extent in the mid-1990s. Short of curbing global greenhouse gas emissions, keeping communities that rely on the Great Lakes livable will require all of these actions and more.

Notes

1. Tom Henry, "Water Crisis Grips Toledo Area," *The Blade*, August 3, 2014, https://www.toledoblade.com/local/2014/08/03/Water-crisis-grips -area/stories/20140803090.
2. "Climate at a Glance," National Climatic Data Center, https://www.ncdc .noaa.gov/cag/regional/time-series/103/pcp/12/12/1895- 2020?base _prd=true&begbaseyear=1901&endbaseyear=2000.
3. "Midwest," U.S. Climate Resilience Toolkit, https://toolkit.climate.gov /regions/Midwest.

4. Samantha Gross, "What Is the Trump Administration's Track Record on the Environment?," Brookings Institution, August 4, 2020, https://www.brookings.edu/policy2020/votervital/what-is-the-trump-administrations-track-record-on-the-environment/.

5. Q. F. Hamlin, A. D. Kendall, S. L. Martin, H. D. Whitenack, J. A. Roush, B. A. Hannah, and D. W. Hyndman, "Quantifying Landscape Nutrient Inputs with Spatially Explicit Nutrient Source Estimate Maps," *Journal of Geophysical Research: Biogeosciences* 125, no. 2 (2020): https://doi.org/10.1029/2019jg005134.

Chapter 5

PFAS "Forever Chemicals" Are Widespread and Threaten Human Health—Here's a Strategy for Protecting the Public

CAROL KWIATKOWSKI

LIKE MANY INVENTIONS, the discovery of Teflon happened by accident. In 1938 chemists from DuPont (now Chemours) were studying refrigerant gases when, much to their surprise, one concoction solidified. Upon investigation, they found it was not only the slipperiest substance they'd ever seen—it was also noncorrosive, extremely stable, and had a high melting point.

In 1954 the revolutionary "nonstick" Teflon pan was introduced. Since then, an entire class of human-made chemicals has evolved: per- and polyfluoroalkyl substances, better known as PFAS. There are upward of 6,000 of these chemicals. Many are used for stain-, grease-, and water-proofing. PFAS are found in clothing, plastic, food packaging, electronics, personal care products, firefighting foams, medical devices, and numerous other products. Over time, though, evidence has slowly built that shows some commonly used PFAS are toxic and may cause cancer. It took 50 years to understand that the happy accident of Teflon's discovery was, in fact, a train wreck.

As a public health analyst, I have studied the harm caused by these chemicals. I am one of hundreds of scientists who have called for a comprehensive, effective plan to manage the entire class of PFAS to protect public health while safer alternatives are developed.

Typically, when the US Environmental Protection Agency assesses chemicals for potential harm, it examines one substance at a time. That approach hasn't worked for PFAS, given the sheer number of them and the fact that manufacturers commonly replace toxic substances with "regrettable substitutes"—similar, lesser-known chemicals that also threaten human health and the environment.

Toxic Chemicals

A class-action lawsuit brought this issue to national attention in 2005. Workers at a Parkersburg, West Virginia, DuPont plant joined with local residents to sue the company for releasing millions of pounds of one of these chemicals, known as PFOA, into the air and the Ohio River. Lawyers discovered that the company had known as far back as 1961 that PFOA could harm the liver.

The suit was ultimately settled in 2017 for US$670 million, after an eight-year study of tens of thousands of people who had been exposed. Based on multiple scientific studies, this review concluded that there was a probable link between exposure to PFOA and six categories of diseases: diagnosed high cholesterol, ulcerative colitis, thyroid disease, testicular cancer, kidney cancer, and pregnancy-induced hypertension.

Over the past two decades, hundreds of peer-reviewed scientific papers have shown that many PFAS are not only toxic—they also don't fully break down in the environment and have accumulated in the bodies of people and animals around the world.[1] Some studies have detected PFAS in 99 percent of people tested.[2] Others have found PFAS in wildlife, including polar bears, dolphins, and seals.[3]

Widespread and Persistent

PFAS are often called "forever chemicals" because they don't fully degrade. They move easily through air and water and can quickly travel long distances and accumulate in sediment, soil, and plants. They have also been found in dust and food, including eggs, meat, milk, fish, fruits, and vegetables.

In the bodies of humans and animals, PFAS concentrate in various organs, tissues, and cells. The US National Toxicology Program and the Centers for Disease Control and Prevention have confirmed a long list of health risks from PFAS, including immunotoxicity, testicular and kidney cancer, liver damage, decreased fertility, and thyroid disease.

Children are even more vulnerable than adults because they can ingest more PFAS relative to their body weight from food and water and through the air. Children also put their hands in their mouths more often, and their metabolic and immune systems are less developed. Studies show that these chemicals harm children by causing kidney dysfunction, delayed puberty, asthma, and altered immune function.[4]

Researchers have also documented that exposure to PFAS reduces the effectiveness of vaccines,[5] which is particularly concerning amid the COVID-19 pandemic.

Widespread in Drinking Water

PFAS have become so ubiquitous in the environment that health experts say it is probably impossible to prevent exposure completely. These substances are released throughout their life cycles, from chemical production to product use and disposal. Up to 80 percent of environmental pollution from common PFAS, such as PFOA, comes from the production of fluoropolymers that use toxic PFAS as processing aids to make products like Teflon.

In 2009 the US Environmental Protection Agency established a health advisory level for PFOA in drinking water of 400 parts per trillion. Health advisories are not binding regulations—they are technical guidelines for state, local, and tribal governments, which are primarily responsible for regulating public water systems. In 2016 the agency dramatically lowered this recommendation to 70 parts per trillion. Some states have set far more protective levels—as low as 8 parts per trillion.[6]

According to the Environmental Working Group, a public health advocacy organization, up to 110 million Americans could be drinking PFAS-contaminated water. Even with the most advanced treatment processes, it is extremely difficult and costly to remove these chemicals from drinking water.[7] And it's impossible to clean up lakes, river systems, or oceans.

Nonetheless, PFAS are largely unregulated by the federal government, although they are gaining increased attention from Congress.

Reducing PFAS Risks at the Source

Given that pollution from PFAS is so ubiquitous and hard to remove, many health experts assert that the only way to address it is by reducing PFAS production and use as much as possible.

Educational campaigns and consumer pressure are making a difference. Many forward-thinking companies, including grocers, clothing manufacturers, and furniture stores, have removed PFAS from products they use and sell. State governments have also stepped in. California banned PFAS in firefighting foams. Maine and Washington have banned PFAS in food packaging. Other states are considering similar measures.

I am part of a group of scientists from universities, non-profit organizations, and government agencies in the United States and Europe that has argued for managing the entire class of PFAS chemicals as a group, instead of one by one. We also support an "essential uses" approach that would restrict their production and use only to products that are critical for health and the proper functioning of society, such as medical devices and safety equipment. And we have recommended developing safer non-PFAS alternatives.

There is an urgent need for innovative solutions to PFAS pollution. Guided by good science, I believe we can effectively manage PFAS to reduce further harm, while researchers find ways to clean up what has already been released.

Notes

1. Philippe Grandjean, "Delayed Discovery, Dissemination, and Decisions on Intervention in Environmental Health: A Case Study on Immunotoxicity of Perfluorinated Alkylate Substances," *Environmental Health* 17, no. 1 (2018): https://doi.org/10.1186/s12940-018-0405-y.
2. Chang Ho Yu, C. David Rikera, Shou-en Lu, and Zhihua Fan, "Biomonitoring of Emerging Contaminants, Perfluoroalkyl and Polyfluoroalkyl Substances (PFAS), in New Jersey Adults in 2016–2018," *International Journal of Hygiene and Environmental Health* 223, no. 1 (2020): 34–44, https://doi.org/10.1016/j.ijheh.2019.10.008.
3. Derek Muir, Rossana Bossi, Pernilla Carlsson, Marlene Evans, Amila De Silva, Crispin Halsall, Cassandra Rauert, et al., "Levels and Trends of Poly- and Perfluoroalkyl Substances in the Arctic Environment—an Update," *Emerging Contaminants* 5 (2019): 240–71, https://doi.org/10.1016/j.emcon.2019.06.002.
4. Kristen Rappazzo, Evan Coffman, and Erin P. Hines, "Exposure to Perfluorinated Alkyl Substances and Health Outcomes in Children: A Systematic Review of the Epidemiologic Literature," *International Journal of Environmental Research and Public Health* 14, no. 7 (2017): https://doi.org/10.3390/ijerph14070691.
5. Philippe Grandjean, Elisabeth Wreford Andersen, Esben Budtz-Jørgensen, Flemming Nielsen, Kåre Mølbak, Pal Weihe, and Carsten Heilmann, "Serum Vaccine Antibody Concentrations in Children Exposed to Perfluorinated Compounds," *JAMA* 307, no. 4 (2021): 391–97, https://doi.org/10.1001/jama.2011.2034.
6. Gloria B. Post, "Recent US State and Federal Drinking Water Guidelines for Per- and Polyfluoroalkyl Substances," *Environmental Toxicology and Chemistry* 40, no. 3 (2021): 550–63, https://doi.org/10.1002/etc.4863.
7. Mohammad Feisal Rahman, Sigrid Peldszus, and William B.Anderson, "Behaviour and Fate of Perfluoroalkyl and Polyfluoroalkyl Substances (PFASs) in Drinking Water Treatment: A Review," *Water Research* 50 (2014): 318–40, https://doi.org/10.1016/j.watres.2013.10.045.

Part II.

**Digging Deeper to
Get More Water**

It's been predicted for many years that water will be the "new oil"—one of the most, if not the most, valued commodities on Earth. Already, it's not hard to see how disruptive a lack of water is to society, as more parts of the world deal with drought and water scarcity. Depleted drinking water supplies undermine local communities, many that are already struggling with public health and social issues such as the COVID-19 pandemic, the opioid crisis, and (in rural communities) soaring farm losses.

As rising temperatures and drought dry up rivers, many are searching for new sources of water and experimenting with novel technologies to harvest new water sources. This includes new technological strategies to recharge diminishing groundwater such as managed aquifer recharge—as Grunes, Seltzer, and Befus relate in their chapter, which emphasizes how old, and vulnerable, much of the world's groundwater really is. Qadir and Smakhtin describe a number of unusual technologies that water-scarce countries are exploiting: fog catching, cloud seeding, desalination, iceberg harvesting, and rainwater harvesting. Better understanding these approaches, including their potential advantages and their limitations, is an important next step.

The need for investment to address water scarcity raises concerns about the rising cost of water and questions of how to keep water affordable and accessible. In some parts of the world, drought and climate change have attracted speculators to the water market. Bruno and Schweizer report on the world's first futures market for water launched in late 2020, which allows speculators to wager on the future price of water.

The good news is that we are not short on innovative governance solutions. In the face of shortage, researchers are proposing new mechanisms to better allocate water and engage stakeholders—like Native American tribes—that historically have been excluded from these conversations in the Colorado River Basin, according to McCool. Others, like Sanderson, Griggs, and Miller-Klugesherz, in their examination of the Ogallala Aquifer's depletion in the US Central Plains, recommend a reexamination of our federal farm policies and point out the need to dismantle structural processes that encourage water waste. There is also a lot to be learned from other places in the world that are experimenting with new management strategies for water and sewer investments, while still protecting the poor and keeping clean water affordable. One such place is Chile, as Cook describes in his chapter.

The chapters of part II span multiple regions to bring attention to water's limited availability and to unsustainable practices around quenching our thirst for water.

Chapter 6

Ancient Groundwater:
Why the Water You're Drinking
May Be Thousands of Years Old

MARISSA GRUNES, ALAN SELTZER, and KEVIN M. BEFUS

COMMUNITIES THAT RELY ON the Colorado River are facing a water crisis. Lake Mead, the river's largest reservoir, has fallen to levels not seen since it was created by the construction of the Hoover Dam roughly a century ago. Arizona and Nevada faced their first-ever mandated water cuts in the summer of 2021, while water was being released from other reservoirs to keep the Colorado River's hydropower plants running. If even the mighty Colorado and its reservoirs are not immune to the heat and drought worsened by climate change, where will the West get its water?

There's one hidden answer: underground.

As rising temperatures and drought dry up rivers and melt mountain glaciers, people depend increasingly on the water beneath their feet. Groundwater resources currently supply drinking water to nearly half the world's population and roughly 40 percent of water used for irrigation globally.[1] What many people don't realize is how old—and how vulnerable—much of that water is.

Most water stored underground has been there for decades, and much of it has sat for hundreds, thousands, or even millions of years. Older groundwater tends to reside deep underground, where it is less easily affected by surface conditions such as drought and pollution. As shallower wells dry out under the pressure of urban development, population growth, and climate change, old groundwater is becoming increasingly important.

Drinking Ancient Groundwater

If you bit into a piece of bread that was 1,000 years old, you'd probably notice.

Water that has been underground for a thousand years can taste different, too. It leaches natural chemicals from the surrounding rock, changing its mineral content. Some natural contaminants linked to groundwater age—like mood-boosting lithium[2]—can have positive effects. Other contaminants, like iron and manganese, can be troublesome. Older groundwater is also sometimes too salty to drink without expensive treatment. This problem can be worse near the coasts: overpump-

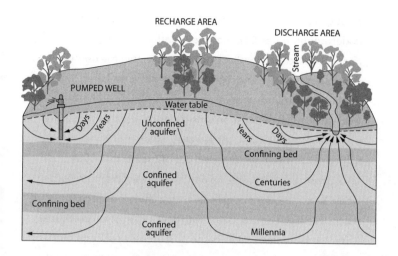

RECHARGE AREA

DISCHARGE AREA

PUMPED WELL

Water table

Days

Years

Unconfined aquifer

Years

Days

Stream

Confining bed

Confined aquifer

Centuries

Confining bed

Confined aquifer

Millennia

ing creates space that can draw seawater into aquifers and contaminate drinking supplies.

Ancient groundwater can take thousands of years to replenish naturally. And, as California saw during its 2011–2017 drought, natural underground storage spaces compress as they empty, so they can't refill to their previous capacity.[3] This compaction in turn causes the land above to crack, buckle, and sink. Yet people today are drilling deeper wells in the West as droughts deplete surface water and farms rely more heavily on groundwater.

What Does It Mean for Water to Be "Old"?

Let's imagine a rainstorm over central California 15,000 years ago. As the storm rolls over what's now San Francisco, most of the rain falls into the Pacific Ocean, where it will eventually evaporate back into the atmosphere. However, some rain also falls into rivers and lakes and over dry land. As that rain seeps through layers of soil, it enters slowly trickling "flowpaths" of underground water. Some of these paths lead deeper and deeper, where water collects in crevices within the bedrock hundreds of meters underground. The water gathered in these underground reserves is in a sense cut off from the active water cycle—at least on timescales relevant to human life.

In California's arid Central Valley, much of the accessible ancient water has been pumped out of the earth, mostly for agriculture.[4] Whereas the natural replenishment timescale would be on the order of millennia, agricultural seepage has partially refilled some aquifers with newer—too often polluted—water.[5] In fact, places like Fresno now actively refill aquifers with clean water (such as treated wastewater or stormwater) in a process known as managed aquifer recharge.

In 2014, midway through its worst drought in modern memory, California became the last western state to pass a law requiring local groundwater sustainability plans. Groundwater may be resilient to heat waves and climate change, but if you use it all, you're in trouble.

One response to water demand? Drill deeper. Yet that answer isn't sustainable.

First, it's expensive: large agricultural companies and lithium-mining firms tend to be the sort of investors that can afford to drill deep enough, while small rural communities can't.

Second, once you pump ancient groundwater, aquifers need time to refill. Flowpaths may be disrupted, choking off a natural water supply to springs, wetlands, and rivers. Meanwhile, the change in pressure underground can destabilize the earth, causing land to sink and even lead to earthquakes.[6]

Third, contamination: While deep, mineral-rich ancient

groundwater is often cleaner and safer to drink than younger, shallower groundwater, overpumping can change that. As water-strapped regions rely more heavily on deep groundwater, overpumping lowers the water table and draws down polluted modern water that can mix with the older water. This mixing causes the water quality to deteriorate, leading to demand for ever-deeper wells.

Reading Climate History in Ancient Groundwater

There are other reasons to care about ancient groundwater. Like actual fossils, extremely old "fossil groundwater" can teach us about the past.

Envision our prehistoric rainstorm again: 15,000 years ago, the climate was quite different from today. Chemicals that dissolved in ancient groundwater are detectable today, opening windows into a past world. Certain dissolved chemicals act as clocks, telling scientists the groundwater's age. For example, we know the rate of decay for dissolved carbon-14 and krypton-18, so we can measure them to calculate when the water last interacted with air.

Younger groundwater that disappeared underground after the 1950s has a unique, man-made chemical signature: high levels of tritium from atomic bomb testing.

Other dissolved chemicals behave like tiny thermometers. Noble gasses like argon and xenon, for instance, dissolve more in cold water than in warm water, along a precisely known temperature curve. Once groundwater is isolated from air, dissolved noble gasses don't do much. As a result, they preserve information about environmental conditions at the time the water first seeped into the subsurface.

The concentrations of noble gasses in fossil groundwater

have provided some of our most reliable estimates of temperature on land during the last ice age. Such findings provide insight into modern climates, including how sensitive Earth's average temperature is to carbon dioxide in the atmosphere. These methods support an important study that found 3.4 degrees Celsius of warming with each doubling of carbon dioxide.[7]

Groundwater's Past and Future

People in some regions, like New England, have been drinking ancient groundwater for years with little danger of exhausting usable supplies. Regular rainfall and varied water sources—including surface water in lakes, rivers, and snowpack—provide alternatives to groundwater and also refill aquifers with new water. If aquifers can keep up with the demand, the water can be used sustainably.

Out west, though, over a century of unmanaged and exorbitant water use means that some of the places most dependent on groundwater—arid regions vulnerable to drought—have squandered the ancient water resources that once existed underground.

A famous precedent for this problem is in the Great Plains. There, the ancient water of the Ogallala Aquifer supplies drinking water and irrigation for millions of people and farms from South Dakota to Texas. If people were to pump this aquifer dry, it would take thousands of years to refill naturally. It is a vital buffer against drought, yet irrigation and water-intensive farming are lowering its water levels at unsustainable rates.

As the planet warms, ancient groundwater is becoming increasingly important—whether flowing from your kitchen tap, irrigating food crops, or offering warnings about Earth's past that can help us prepare for an uncertain future.

Notes

1. Petra Döll, Heike Hoffmann-Dobrev, Felix T. Portmann, Stefan Siebert, Annette Eicker, Matthew Rodell, Gil Strassberg, and B. R. Scanlon, "Impact of Water Withdrawals from Groundwater and Surface Water on Continental Water Storage Variations," *Journal of Geodynamics* 59–60 (2012): 143–56, https://doi.org/10.1016/j.jog.2011.05.001.

2. Anjum Memon, Imogen Rogers, Sophie M. D. D. Fitzsimmons, Ben Carter, Rebecca Strawbridge, Diego Hidalgo-Mazzei, and Allan H. Young, "Association between Naturally Occurring Lithium in Drinking Water and Suicide Rates: Systematic Review and Meta-analysis of Ecological Studies," *British Journal of Psychiatry* 217, no. 6 (2020): 667–78, https://doi.org/10.1192/bjp.2020.128.

3. Ryan G. Smith, Rosemary Knight, Jingyi Chen, J. A. Reeves, H. A. Zebker, Tom Farr, and Zhen Liu, "Estimating the Permanent Loss of Groundwater Storage in the Southern San Joaquin Valley, California," *Water Resources Research* 53, no. 3 (2017): 2133–48, https://doi.org/10.1002/2016wr019861.

4. Bryant C. Jurgens, John Karl Böhlke, Leon J. Kauffman, Kenneth Belitz, and Bradley K. Esser, "A Partial Exponential Lumped Parameter Model to Evaluate Groundwater Age Distributions and Nitrate Trends in Long-Screened Wells," *Journal of Hydrology* 543 (2016): 109–26, https://doi.org/10.1016/j.jhydrol.2016.05.011.

5. Alan M. Seltzer, David V. Bekaert, Peter H. Barry, Kathryn E. Durkin, Emily K. Mace, Craig E. Aalseth, Jake C. Zappala, Peter Mueller, Bryant Jurgens, and Justin T. Kulongoski, "Groundwater Residence Time Estimates Obscured by Anthropogenic Carbonate," *Science Advances* 7, no. 17 (2017): https://doi.org/10.1126/sciadv.abf3503.

6. Nadav Wetzler, Eyal Shalev, Thomas Göbel, Falk Amelung, Ittai Kurzon, Vladimir Lyakhovsky, and Emily E. Brodsky, "Earthquake Swarms Triggered by Groundwater Extraction Near the Dead Sea Fault," *Geophysical Research Letters* 46, no. 14 (2019): 8056–63, https://doi.org/10.1029/2019gl083491.

7. Jessica E. Tierney, Jiang Zhu, Jonathan King, Steven B. Malevich, Gregory J. Hakim, and Christopher J. Poulsen, "Glacial Cooling and Climate Sensitivity Revisited," *Nature* 584, no. 7822 (2020): 569–73, https://doi.org/10.1038/s41586-020-2617-x.

Chapter 7

As Climate Change Parches the Southwest, Here's a Better Way to Share Water from the Shrinking Colorado River

DANIEL CRAIG McCOOL

THE COLORADO RIVER is a vital lifeline for the arid US Southwest. It supplies water to 7 states, Mexico, 29 Indian reservations, and millions of acres of irrigated farmland. The river and its tributaries support 16 million jobs and provide drinking water to Denver, Salt Lake City, Albuquerque, Las Vegas, Los Angeles, San Diego, Phoenix, and Tucson—in all, 40 million people.

The river also courses through several of the world's most

iconic national parks, including the Grand Canyon in Arizona and Canyonlands in Utah. Today millions of people visit the Colorado River Basin to fish, boat, and explore.

Southwestern states, tribes, and Mexico share the Colorado's water under the century-old 1922 Colorado Compact and updates to it.[1] But today, because of climate change and rapid development, there is an enormous gap between the amount of water the compact allocates to parties and the amount that is actually in the river. With users facing unprecedented water shortages, the compact is hopelessly inadequate to deal with current and future realities.

I have studied water resource development for 35 years and written extensively about Native American water rights and the future of America's rivers. As I see it, the compact rests on three fundamental errors that now plague efforts to develop a new vision for the region. I believe the most productive way forward is for states and tribes to negotiate a new agreement that reflects 21st-century realities.

Flawed Data and Allocations

The compact commissioners made two fatal blunders when they allocated water in 1922. First, they appraised the river's volume based on inaccurate data that wildly overestimated it. Actual annual historical flows were far below what was needed to satisfy the dictates of the compact. There is evidence that the commissioners did this purposefully: reaching an agreement was easier if there was more water to go around. This strategy guaranteed that the compact would allocate more water than was actually in the river, a situation now referred to as the "structural deficit."

Second, the compact allocated water in fixed amounts

rather than percentages of the river's actual flow. That approach would be viable if river flow were constant and the agreement were based on sound science. But the Colorado's flow is highly variable. The compact divided the river artificially into an Upper Basin (Wyoming, Colorado, Utah, and New Mexico) and a Lower Basin (Arizona, Nevada, and California) and allocated 7.5 million acre-feet of water to each basin. An acre-foot is enough water to cover an acre of land to a depth of a foot, or about 325,000 gallons.

In 1944 a treaty allocated an additional 1.5 million acre-feet to Mexico, for a total of 16.5 million acre-feet. However, actual flow has typically been below that amount. River volume at the time of the compact was about 18 million acre-feet per year, but the 20th-century average was closer to 14.8 million acre-feet. And then things got much worse.

In the past 20 years, climate change has further reduced the Colorado's volume.[2] A "megadrought" has reduced flows by nearly 20 percent, and studies predict that it will fall 20 percent to 35 percent or more by mid-century. In late August 2021, Lake Mead, the nation's largest reservoir, was just 35 percent full. Lake Powell, the second-largest US reservoir, was less than 30 percent full. That month, the Bureau of Reclamation declared an official shortage, which will force Arizona, Nevada, and Mexico to make significant cuts in water use. In short, the original fixed allocations are no longer anchored in reality.

In my view, a much better approach would be to allocate water among the states and tribes in percentages, based on a five-year rolling average that would change as the river's flow changes. Without such a shift, the compact will merely perpetuate a hydrological fallacy that leads water users to claim water that does not actually exist.

No Native Participation

Beyond these errors, the compact also rests on a fundamental injustice. The 30 tribal nations in the Colorado River Basin are the river's original users, and their reservations encompass huge swaths of land. But they were completely left out of the 1922 allocations. Compact commissioners, whose views reflected the overt racism of that era, assumed Native peoples did not deserve their own allocation.[3] Making matters worse, nearly every statute, compact, and regulation promulgated since 1922—a body of rules known collectively as the Law of the River—has either ignored or marginalized Native water users.[4] Many tribes, scholars, and advocacy groups view this as an injustice of monumental proportions.

Tribes have gone to court to claim a share of the Colorado's water and have won significant victories, beginning with the landmark 1963 *Arizona v. California* ruling, in which the US Supreme Court recognized water rights for five Indian reservations in the Colorado River Basin. The tribes continued to press their claims through numerous negotiated settlements,[5] starting in 1978 and continuing to this day. They now have rights to over 2 million acre-feet of water in the Lower Basin and 1.1 million acre-feet in the Upper Basin. And 12 tribes have unresolved claims that could total up to 405,000 acre-feet.

Currently, however, tribes are not drawing all of their water because they don't have the pipelines and other infrastructure that they need to divert and use it. This allows non-Indian communities downstream to use the surplus water, without payment in most cases. I believe a new compact should include tribes as equal partners with states and give them meaningful and significant roles in all future negotiations and policy making in the basin.

A New Vision

The compact states are now renegotiating interim river management guidelines that were first adopted in 2007. This process must be completed by 2026 when that agreement expires. I see these discussions as an excellent opportunity to discard the compact's unworkable provisions and negotiate a new agreement that responds to the unprecedented challenges now affecting the Southwest. As I see it, an agreement negotiated by and for white men, based on egregiously erroneous data, in an age when people drove Model T cars cannot possibly serve as the foundation for a dramatically different future.

In my view, the 1922 compact is now an albatross that can only inhibit innovation. Eliminating fixed rights to water that doesn't actually exist could spur members to negotiate a new, science-based agreement that is fairer, more inclusive, and more efficient and sustainable.

Notes

1. Jason Anthony Robison, "The Colorado River Revisited," *University of Colorado Law Review* 88 (2016): 475–570, https://doi.org/10.2139/ssrn.2727279.
2. Bradley Udall and Jonathan Overpeck, "The Twenty-First Century Colorado River Hot Drought and Implications for the Future," *Water Resources Research* 53, no. 3 (2017): 2404–18, https://doi.org/10.1002/2016wr019638.
3. Jason Robison, Daniel McCool, and Thomas Minckley, eds., *Vision and Place: John Wesley Powell and Reimagining the Colorado River Basin* (Berkeley: University of California Press, 2020).
4. Daniel McCool, Sue Jackson, Jason Anthony Robison, Kelsey Leonard, and Barbara A. Cosens, "Indigenous Water Justice," *Lewis & Clark Law Review* 22 (2018): 841–921, https://doi.org/10.2139/ssrn.3013470.
5. Barbara Cosens and Judith V. Royster, eds., *The Future of Indian and Federal Reserved Water Rights: The Winters Centennial* (Albuquerque: University of New Mexico Press, 2012).

Chapter 8

Farmers Are Depleting
the Ogallala Aquifer Because
the Government Pays Them to Do It

MATTHEW R. SANDERSON, BURKE W. GRIGGS,
and JACOB A. MILLER-KLUGESHERZ*

A SLOW-MOVING CRISIS threatens the US Central Plains, which grow a quarter of the nation's crops. Underground, the region's lifeblood—water—is disappearing, placing one of the world's major food-producing regions at risk. The Ogallala–High Plains Aquifer is one of the world's largest groundwater sources, extending from South Dakota down through the

---

*Stephen Lauer and Vivian Aranda-Hughes, former doctoral students at Kansas State University, contributed to several of the studies cited in this chapter.

Texas Panhandle across portions of eight states.[1] Its water supports US$35 billion in crop production each year.[2]

But farmers are pulling water out of the Ogallala faster than rain and snow can recharge it. Between 1900 and 2008 they drained some 89 trillion gallons from the aquifer—equivalent to two-thirds of Lake Erie. Depletion is threatening drinking water supplies and undermining local communities already struggling with the COVID-19 pandemic, the opioid crisis, hospital closures, soaring farm losses, and rising suicide rates.

In Kansas, "Day Zero"—the day wells run dry—has arrived for about 30 percent of the aquifer. Within 50 years, the entire aquifer is expected be 70 percent depleted.[3] Some observers blame this situation on periodic drought. Others point to farmers, since irrigation accounts for 90 percent of Ogallala groundwater withdrawals. But our research, which focuses on social and legal aspects of water use in agricultural communities, shows that farmers are draining the Ogallala because state and federal policies encourage them to do it.

A Production Treadmill

The fates and fortunes of farmers tend to vary year to year. When prices for their products are rising, and production levels are high, farmers on the Central Plains generally appear to be in good financial shape. And vice versa: when farm prices are declining, or production levels are decreasing, farmers tend to fare less well economically.

But regardless of trends in prices and production, these

figures hide massive government payments to farmers. Billions of dollars in federal funds are transferred to farmers year in and year out in the form of price supports and crop insurance, to name two prominent examples. Our research finds that subsidies put farmers on a treadmill—working harder to produce more while draining more of the resource that supports their livelihood.[4] Government payments create a vicious cycle of overproduction that intensifies water use. Subsidies encourage farmers to expand and buy expensive equipment to irrigate larger areas.

With low market prices for many crops, production does not cover expenses on most farms. To stay afloat, many farmers buy or lease more acres. Growing larger amounts floods the market, further reducing crop prices and farm incomes. Subsidies support this cycle. Few benefit, especially small and midsized operations. In a 2019 study of the region's 234 counties from 1980 to 2010, we found that larger irrigated acreage failed to increase incomes or improve education or health outcomes for residents.[5]

Focus on Policy, Not Farmers

Four decades of federal, state, and local conservation efforts have mainly targeted individual farmers, providing ways for them to voluntarily reduce water use or adopt more water-efficient technologies. While these initiatives are important, they haven't stemmed the aquifer's decline. In our view, what the Ogallala Aquifer region really needs is policy change.

A lot can be done at the federal level, but the first principle should be "Do no harm." Whenever federal agencies have tried to regulate groundwater, the backlash has been swift and intense, with farm states' congressional representatives repudiating federal jurisdiction over groundwater. Nor should Congress propose to eliminate agricultural subsidies, as some environmental organizations and free-market advocates have proposed. Given the thin margins of farming and long-standing political realities, federal support is simply part of modern production agriculture.

With these cautions in mind, three initiatives could help ease pressure on farmers to keep expanding production. The US Department of Agriculture's Conservation Reserve Program pays farmers to allow environmentally sensitive farmland to lie fallow for at least 10 years. With new provisions, the program could reduce water use by prohibiting expansion of irrigated acreage, permanently retiring marginal lands, and linking subsidies to production of less water-intensive crops.

These initiatives could be implemented through the

federal farm bill, which also sets funding levels for nonfarm subsidies such as the Supplemental Nutrition Assistance Program, or SNAP. Payments from SNAP, which increase needy families' food budgets, are an important tool for addressing poverty. Increasing these payments and adding financial assistance to local communities could offset lower tax revenues that result from farming less acreage.

Amending federal farm credit rates could also slow the treadmill. Generous terms promote borrowing for irrigation equipment; in order to pay that debt, borrowers farm more land. Offering lower rates for equipment that reduces water use and withholding loans for standard, wasteful equipment could nudge farmers toward conservation.

The most powerful tool is the tax code. Currently, farmers receive deductions for declining groundwater levels and can write off depreciation of irrigation equipment. Replacing these perks with a tax credit for stabilizing groundwater and substituting a depreciation schedule favoring more efficient irrigation equipment could provide strong incentives to conserve water.

Rewriting State Water Laws

Water rights are mostly determined by state law, so reforming state water policies is crucial. Case law demonstrates that simply owning water rights does not grant the legal right to waste water. For more than a century, courts have upheld state restrictions on waste, with rulings that allow for adaptation by modifying the definitions of beneficial use and waste over time. Using these precedents, state water agencies could designate thirsty crops, such as rice, cotton, or corn, as wasteful in certain regions. Regulations preventing unreasonable water use are not unconstitutional.

Allowing farmers some flexibility will maximize profits, as long as they stabilize overall water use. If they irrigate less—or not at all—in years with low market prices, rules could allow more irrigation in better years. Ultimately, many farmers and their bankers are willing to exchange lower annual yields for a longer water supply.

As our research has shown, the vast majority of farmers in the region want to save groundwater.[6] They will need help from policy makers to do it. Forty years is long enough to learn that the Ogallala Aquifer's decline is not driven by weather or by individual farmers' preferences. Depletion is a structural problem embedded in agricultural policies. Groundwater depletion is a policy choice made by federal, state, and local officials.

Notes

1. Matthew R. Sanderson and R. Scott Frey, "Structural Impediments to Sustainable Groundwater Management in the High Plains Aquifer of Western Kansas," *Agriculture and Human Values* 32, no. 3 (2014): 401–17, https://doi.org/10.1007/s10460-014-9567-6.

2. Bruno Basso, Anthony D. Kendall, and David W. Hyndman, "The Future of Agriculture over the Ogallala Aquifer: Solutions to Grow Crops More Efficiently with Limited Water," *Earth's Future* 1, no. 1 (2013): 39–41, https://doi.org/10.1002/2013ef000107.

3. David R. Steward, Paul J. Bruss, Xiaoying Yang, Scott A. Staggenborg, Stephen M. Welch, and Michael D. Apley, "Tapping Unsustainable Groundwater Stores for Agricultural Production in the High Plains Aquifer of Kansas, Projections to 2110," *Proceedings of the National Academy of Sciences* 110, no. 37 (2013): E3477–86, https://doi.org/10.1073/pnas.1220351110.

4. Matthew R. Sanderson and Vivian Hughes, "Race to the Bottom (of the Well): Groundwater in an Agricultural Production Treadmill," *Social Problems* 66, no. 3 (2018): 392–410, https://doi.org/10.1093/socpro/spy011.

5. Stephen Lauer and Matthew Sanderson, "Irrigated Agriculture and Human Development: A County-Level Analysis, 1980–2010," *Environment, Development and Sustainability* 22, no. 5 (2019): 4407–23, https://doi.org/10.1007/s10668-019-00390-9.

6. Stephen Lauer and Matthew R. Sanderson, "Producer Attitudes toward Groundwater Conservation in the U.S. Agallala–High Plains," *Groundwater* 58, no. 4 (2019): 674–80, https://doi.org/10.1111/gwat.12940.

Chapter 9

Millions of Americans Struggle to Pay Their Water Bills—Here's How a National Water Aid Program Could Work

JOSEPH COOK

RUNNING WATER AND INDOOR PLUMBING are so central to modern life that most Americans take them from granted. But these services aren't free, and millions struggle to afford them. A 2019 survey found that US households in the bottom fifth of the economy spent 12.4 percent of their disposable income on water and sewer services.[1] News reports suggest that for low-income households, this burden increased during the COVID-19 pandemic.

Since 1981, the federal government has helped low-

income households with their energy costs through the
Low-Income Home Energy Assistance Program. But there
had not been a national water aid program until Congress
created a temporary Low-Income Household Water Assistance Program as part of the COVID-19 response. The Build
Back Better Act, passed in 2021, includes US\$225 million for
grants to states and tribes to help reduce the cost of water
services for low-income households.[2]

As an economist specializing in environmental and natural resource issues, I'm encouraged to see this idea gaining
support. But I also know from analyzing efforts at the local
level that these programs may be ineffective if they aren't well
designed. I believe the United States can learn lessons from
Chile, which has run an effective national water assistance
program for 30 years.

Flaws in US Local Aid Programs

I have studied water and sewer customer assistance programs
around the world and developed a database of examples run
by US utilities in cities including Seattle, Philadelphia, and Baltimore. Although there are hundreds of these programs, three
major problems undercut their effectiveness.

First, because utilities have to fund their assistance
programs from their own budgets, they typically charge
"non-poor" customers higher rates and use those payments
to subsidize low-income customers. State regulations often
forbid this, forcing utilities in those states to rely on voluntary

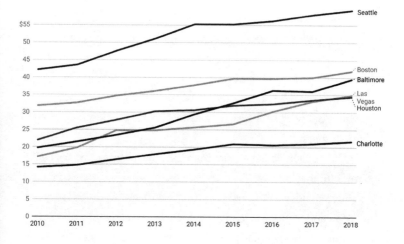

donation programs to fund assistance. Second, in areas with high poverty, too many customers need help, and there are not enough non-poor customers to foot the bill. Third, smaller and less well-funded utilities often do not have the administrative capacity or expertise to design and implement their own customer assistance programs.

These challenges have spurred politicians and policy experts to call for a federal program—a step that the Environmental Protection Agency's National Drinking Water Advisory Council recommended back in 2003.

Learning from Chile's Experience

Agencies such as the World Bank often cite Chile's water aid program as a model. Here's how it works.

The program aims to ensure that households don't pay more than 3 percent of their income for receiving a quantity of water to meet their essential needs. There is no consensus among experts on what this "lifeline" quantity of water should be, but Chile sets it at 15 cubic meters per month—about 4,000 gallons. Eligible customers apply to their city government every three years. Once enrolled, they immediately see reductions in their bills, based on their poverty levels, for that first 15 cubic meters of water use. Each month, the water utility bills the city for subsidies it has provided to poor customers.

If households use more than 15 cubic meters of water per month, they pay unsubsidized prices for whatever they use above that level. This gives everyone an incentive to fix leaky pipes and appliances and conserve water. Regulators who set water prices are not involved in running the subsidy system or determining subsidy levels.

In the late 1990s, Chile launched a major expansion of its sewage treatment plants. Water utilities raised their rates by 34 percent to 142 percent between 1998 and 2015 to pay for this initiative.[3] Because these rate increases outpaced growth in income, subsidies grew by 54 percent over the same time period. The takeaway: Chile found a way to pay for water and sewer investment while still protecting the poor.

How a US Water Aid Program Might Work

If the United States creates a national water aid program, key questions will include who is eligible and how much water is an "essential" quantity for households. The Environmental Protection Agency estimates that an average US household uses approximately 9,000 gallons per month, but one-third of this is for gardens and lawns. Reliable national data on US household water usage is nearly nonexistent, and there is no estimate of how much water low-income households use. Program managers would need to collect information on utility water and sewer pricing structures, put it in a database, and couple it with census data to estimate the number of eligible households in each state.

To estimate what a program like Chile's might cost here, my team at Washington State University compiled a database of water and sewer rates as of December 2019. We included all US cities with populations over 100,000, at least two cities per state, and made assumptions about rates for smaller cities and towns. We estimate that a program covering the full cost of 4,500 gallons of water per month for households at or below the poverty line would cost approximately $11.2 billion annually if 70 percent of eligible households participate. In total, we estimate that 11.8 million households would receive an average subsidy of US$67 per month.

Our project website includes a calculator tool to estimate the annual federal cost based on different assumptions about eligibility, participation, the essential water quantity, and the percentage discount on water bills.

Another Approach: Add Money to SNAP Payments

Public policy scholar Manny Teodoro has suggested another way to deliver water aid: topping up support that people receive to buy healthy food through the Supplemental Nutrition Assistance Program, or SNAP. This idea builds on a well-known program with a long track record. Low-income households would not have to file new paperwork to receive benefits. Delivering water aid this way could help renters, whose water costs often are rolled into their rent, and rural residents who use well water and have to pay for water treatment and maintenance costs out of pocket. It would place less of an administrative burden on the large number of small US water systems serving fewer than 500 people. And it could be quickly implemented by adding water providers as approved vendors for payments through an electronic benefit transfer card.

Eligibility for SNAP is set at 138 percent of the poverty line, and an estimated 84 percent of eligible households participate.[4] With these parameters, we estimate that a program covering 100 percent of the cost of 4,500 gallons of water per month would cost $17 billion annually. The main weakness of this approach is that water and sewer rates vary across the country, so it risks providing too much or too little assistance to low-income households depending on where they live.

Getting Water Prices Right for Everyone

Access to a safe and affordable water supply and sewer services is codified in the United Nations' human right to water and sanitation. The United States is a wealthy country, and my research group's estimates show that the cost of a targeted program to help the poor pay their bills is reasonable. Without federal funding, poor and marginalized households will continue to fall behind on their bills and experience the indignity and health risks of having their water turned off.

At the same time, the United States needs to make major investments in its water and sewer infrastructure and manage the effects of drought and climate change. Economists broadly agree that water should be more expensive in many places to give local governments and ratepayers incentive to conserve and plan for a water-scarce future.[5] I believe Chile's experience shows how a national program can preserve this signal while directing most of its water sector subsidies toward protecting the poor.

Notes

1. Manuel P. Teodoro and Robin Rose Saywitz, "A Snapshot of Water and Sewer Affordability in the United States, 2019," *Journal: American Water Works Association* 112, no. 8 (2020): 10–19, https://mannyteodoro.com/wp-content/uploads/TeodoroSwaywitz-JAWWA-2020-Affordability-Snapshot.pdf.
2. *Build Back Better Act—Rules Committee Print 117–18 Section-by-Section*, House Committee on Rules, https://rules.house.gov/sites/democrats.rules.house.gov/files/Section_by_Section_BBB.pdf.
3. Dante Contreras, Andrés Gómez-Lobo, and Isidora Palma, "Revisiting the Distributional Impacts of Water Subsidy Policy in Chile: A Historical Analysis from 1998–2015," *Water Policy* 20, no. 6 (2018): 1208–26. https://doi.org/10.2166/wp.2018.073.
4. United States Department of Agriculture, *Reaching Those in Need: Estimates of State Supplemental Nutrition Assistance Program Participation Rates in 2017*, 2020, https://fns-prod.azureedge.net/sites/default/files/resource-files/Reaching2017-1.pdf.
5. Eduardo Porter, "The Risks of Cheap Water," *New York Times*, October 14, 2014, https://www.nytimes.com/2014/10/15/business/economy/the-price-of-water-is-too-low.html.

Chapter 10

Five Unusual Technologies
for Harvesting Water in Dry Areas

MANZOOR QADIR and VLADIMIR SMAKHTIN

WATER SCARCITY is among the top five global risks affecting people's well-being. In water-scarce countries, the situation is grim. Conventional sources like snowfall, rainfall, river runoff, and easily accessible groundwater are being affected by climate change, and supplies are shrinking as demand grows. In these countries, sufficient water supply is a critical challenge to sustainable development and a potential cause of social unrest and conflict. Water scarcity also impacts traditional seasonal human migration routes and, together with other water insecurity factors, could reshape migration patterns.

Water-scarce countries need a fundamental change in planning and management.

We are looking at how to do this, through the creative exploitation of unconventional water resources. From Earth's seabed to its upper atmosphere, we have a variety of water resources that can be tapped. But making the most of these requires a diverse range of technological interventions and innovations.

Catching Fog

Water embedded in fog is increasingly seen as a source of drinking water in dry areas where fog is intense and happens regularly. Fog can be collected using a vertical mesh that intercepts the droplet stream. The captured water then runs down into a water collection, storage, and distribution system.

Different types of screen materials can be used in fog collectors, like aluminum, plastic, plexiglass, and alloy. The success of a system like this depends on the geography and topography, which need to be conducive to optimal fog interception.[1] But this could work in dry mountainous and coastal regions.

With active engagement of local communities and technical support from local institutions, fog water harvesting is a low-maintenance option and a green technology to supply drinking water.[2] Fog water collection projects have been implemented in different parts of world, including Morocco, Chile, Eritrea, Israel, and Oman.

Seeding Clouds

Under the right weather conditions, rain enhancement through cloud seeding has the potential to increase the volume of water harvesting from air. This technology involves dispersing small particles into clouds or in their vicinity. These particles act as a starting point for raindrops or ice crystals, promoting their formation. In turn, this makes it more likely to rain or snow.

Application of cloud seeding technology in different countries has shown that precipitation can be increased by up to 20 percent of the annual norm, depending on the available cloud resources and types, cloud water content, and base temperature.[3] As only up to 10 percent of total cloud water content is released to the ground as precipitation, there is a huge potential for rain enhancement technologies to increase rainfall in dry areas.[4]

Minimizing Evaporation

Because dry areas receive small amounts of rainfall, micro-catchment rainwater harvesting may help in capturing rainwater on the ground, where it would otherwise evaporate.

There are two major types of micro-catchment rainwater harvesting systems. One is water harvesting via rooftop systems, where runoff is collected and stored in tanks or similar devices.[5] This water is used domestically or for livestock watering. The second is water harvesting for agriculture, which involves collecting the rainwater that runs off a catchment area in a small reservoir or in the root zone of a cultivated area.[6] The catchment surface may be natural or treated with a material that stops the soil from absorbing water, especially in areas with sandy soils. Because of the intermittent nature of runoff, it is necessary to store the maximum amount of rainwater during the rainy season so that it can be used later.

Desalinating Seawater

The process of desalination removes salt from seawater or brackish groundwater to make it drinkable.[7] This allows us to gather water beyond what is available from the water cycle, providing a climate-independent and steady supply of high-quality water.

Seawater desalination has been growing faster because of advances in membrane technology and material science. These advances are projected to cause a significant decrease in production costs by 2030. More places are expected to become reliant on desalinated water because of its falling costs and the rising costs of conventional water resources. While, at present, desalination provides approximately 10 percent of the municipal water supply of urban coastal centers worldwide, by the year 2030, this is expected to reach 25 percent.

Harvesting Icebergs

Towing an iceberg from one of the polar ice caps to a water-scarce country may not seem like a practical solution to water shortages, but scientists, scholars, and politicians are considering iceberg harvesting as a potential freshwater source.

Moving an iceberg across the ocean is technically possible, based on a theoretical four-part process. It would require locating a suitable source and supply, calculating the necessary towing power requirements, accurately predicting melting in transit, and estimating the economic feasibility of the entire endeavor. Countries like the United Arab Emirates and South Africa are considering iceberg towing as an option to narrow gaps in their water demand and supply.

Water and climate change are interconnected, so climate change increases the likelihood of extreme droughts in dry areas. Harnessing the potential of unconventional water resources can increase the resilience of water-scarce communities against climate change, while diversifying water supply resources. We need to identify and promote functional systems of unconventional water resources that are environmentally feasible, economically viable, and supportive of water-related sustainable development, as projected in the 2030 agenda for the United Nations Sustainable Development Goals and for the future beyond.

Notes

1. Ayman F. Batisha, "Feasibility and Sustainability of Fog Harvesting," *Sustainability of Water Quality and Ecology* 6 (2015): 1–10. https://doi.org/10.1016/j.swaqe.2015.01.002.

2. Manzoor Qadir, Gabriela C. Jiménez, Rebecca L. Farnum, Leslie L. Dodson, and Vladimir Smakhtin, "Fog Water Collection: Challenges beyond Technology," *Water* 10, no. 4 (2018): https://doi.org/10.3390/w10040372.

3. A. I. Flossmann, M. Manton, A. Abshaev, R. Bruintjes, M. Murakami, T. Prabhakaran, and Z. Yao, "Review of Advances in Precipitation Enhancement Research," *Bulletin of the American Meteorological Society* 100, no. 8 (2019): 1465–80, https://doi.org/10.1175/bams-d-18-0160.1.

4. M. T. Abshaev, A. M. Abshaev, A. M., Malkarova, and Z. Y. Mizieva, "Radar Estimation of Water Content in Cumulonimbus Clouds," *Izvestiya, Atmospheric and Oceanic Physics* 45, no. 6 (2009): 731–36, https://doi.org/10.1134/s0001433809060061.

5. Theib Y. Oweis, "Rainwater Harvesting for Restoring Degraded Dry Agro-pastoral Ecosystems: A Conceptual Review of Opportunities and Constraints in a Changing Climate," *Environmental Reviews* 25, no. 2 (2017): 135–49, https://doi.org/10.1139/er-2016-0069.

6. Russell B. Thomas, Mary Jo Kirisits, Dennis J. Lye, and Kerry A. Kinneya, "Rainwater Harvesting in the United States: A Survey of Common System Practices," *Journal of Cleaner Production* 75 (2014): 166–73, https://doi.org/10.1016/j.jclepro.2014.03.073.

7. Edward Jones, Manzoor Qadir, Michelle T. H. van Vliet, Vladimir Smakhtin, and Seong-mu Kang, "The State of Desalination and Brine Production: A Global Outlook," *Science of the Total Environment* 657 (2019): 1343–56, https://doi.org/10.1016/j.scitotenv.2018.12.076.

Chapter 11

Why Wall Street Investors' Trading of California Water Futures Is Nothing to Fear— and Unlikely to Work Anyway

ELLEN BRUNO and HEIDI SCHWEIZE

WATER IS ONE OF THE WORLD'S most vital resources. So is there reason to freak out now that profit-hungry hedge funds and other investors can trade water futures like barrels of oil or shares of Apple?

That's exactly what CME Group did in California when it launched the world's first futures market for water in December 2020. Put simply, a futures market lets people place bets on the future price of water. Some people worry that Wall Street's involvement in trading water will disenfranchise the water rights of rural communities and lead to more scarcity of an already dwindling resource, thus driving up the price everyone pays.[1]

As researchers who study commodity markets and the economics of water resources, we believe there are many benefits of a well-functioning water futures market, especially as climate change makes the amount available for use increasingly hard to predict. The market's main purpose, after all, is to provide protection for California water users, such as farmers and cities, against fluctuations in prices. While there are real risks, we think they're misunderstood and overblown. And anyway, very few are actually trading water futures.

Futures 101

Farmers and suppliers have used futures markets to manage the risk associated with the changing market prices of commodities for centuries. Such trading wasn't standardized until 1848, when the Chicago Board of Trade became the world's first futures exchange and began trading corn, wheat, and soybean futures.[2] Today there are futures markets for many types of assets, from commodities like coffee and crude oil to currencies such as the dollar and yen.

Here's a typical example. Let's say the price of soybeans is US$12 a bushel. In the spring, a soybean farmer might sell a September futures contract for US$12 a bushel that obligates

her to deliver a certain amount of soybeans in the fall and receive the agreed-upon price in return—regardless of how the actual "spot" price changes. Given all the uncertainties of farming, this makes it easier to plan.

The water futures market works similarly, except there's no physical exchange of the asset—that is, no water actually changes hands. So, someone who needs water, such as a farmer, buys a contract to purchase water in three months at US$500 per acre-foot. If in three months' time a drought drives the spot price up to US$550, the seller of the contract pays the farmer the difference: US$50 per acre-foot. If a sudden increase in the supply of water pushes the price lower to US$450, the farmer must pay the difference. But for the farmer, in either scenario, when she actually buys water for her crops, she pays just US$500 per acre-foot.

The Important Role of Speculators

One of the more unnerving aspects of trading futures for basic resources like food or water on an open exchange is that anyone can participate and speculate on the resource's future price—anyone from hedge funds to amateur traders—not just the ones who actually use it. The worry is that they could affect either the actual price of water or its supply.[3] Understanding why this shouldn't be a concern comes down to zeroing in on the role of hedgers and speculators.

Farmers, power plant operators, cities, and others that rely on water to conduct business can use the futures market like an insurance policy to hedge their risks. A farmer may not want to worry about the price of water increasing in the summer, so she hedges that risk by buying a futures contract that locks in a price. Or a water district that sells water to commercial users may want to hedge against a drop in price, so it sells a contract. The futures market allows these water users to trade away risk.

But this wouldn't be easy without the participation of lots of investors and other speculators to grease the wheels by agreeing to be the buyer or seller of any given contract. In finance, this is called creating "liquidity," and it allows traders to buy and sell this risk with one another. Although purely speculative trading activity is often blamed for market booms and busts, commodity market researchers have found not evidence of this.[4]

Possible Problems

However, the hedging benefits of water futures are contingent on the market functioning properly. And there are several key differences between California water futures and more well-established markets. For one thing, futures markets work best when the price of the underlying asset is well known. Cash settlement of futures contracts on agricultural products like hogs and cattle, for example, relies on price indices that are calculated using publicly available, government-collected data.[5]

There's much less visibility for the price of water. The futures market is based on the Nasdaq Veles California Water Index, which is constructed using confidential transaction data collected by a private company. The lack of transparency over its construction and the underlying data will limit how much confidence market participants can have in the integrity of its values. Furthermore, the complexity of water use and policy means that the true extent of water's scarcity may not be well captured by the index, which would make farmers and other hedgers less likely to participate.

Another challenge is that many of the potential users of a water futures market, such as municipalities and government agencies, may have little experience in the risky world of futures trading. Trading in futures markets requires daily settlement. That means, at the end of every day, the day's losers must pay the winners, so participants must always have cash available. As a result, a public agency could be on the hook for tens of thousands of dollars if it's on the losing end of a contract whose price changes dramatically.

Finally, the weather-dependent nature of water—it either rains or it doesn't—makes it hard to make meaningful predictions about its future price. That makes a futures market seem more like a random roll of the dice than rational speculation. As a result, it's unclear whether there will be enough investor interest to make it work.

A Sleepy Start

These challenges may explain why there was so little activity on the water futures market in its first year. During that period, the number of open contracts on the exchange did not exceed 60 contracts in a single day, a fraction compared with other commodities that have been given cash-settled futures markets in recent years. For example, futures of block cheese, which began trading in January 2020, were averaging more than 400 open contracts a day just two months after launch. Pork cutouts, which are all the various cuts of a whole pig, began trading only a month before California water and were averaging more than 1,500 open contracts a day a few months after launching. Even at the peak, 100 water futures contracts represent 1,000 acre-feet, a trivial share of California's total water use, which can exceed 100 million acre-feet in a wet year.

So even though such a futures market for water could be useful—especially in the face of climate change—we believe the initial effort in California to create water futures is unlikely to take off.

Notes

1. Ben Ryder Howe, "Wall Street Eyes Billions in the Colorado's Water," *New York Times*, January 3, 2021.
2. Timeline of CME Achievements" CME Group, https://www.cmegroup .com/company/history/timeline-of-achievements.html.
3. "Water: Futures Market Invites Speculators, Challenges Basic Human Rights—UN Expert," ReliefWeb, December 11, 2020, https://reliefweb.int /report/world/water-futures-market-invites-speculators-challenges -basic-human-rights-un-expert.
4. Raymond P. H. Fishe and Aaron Smith, "Do Speculators Drive Commodity Prices Away from Supply and Demand Fundamentals?," *Journal of Commodity Markets* 15 (2019): 100078, https://doi.org/10.1016 /j.jcomm.2018.09.006.

5. Colin A. Carter and Sandeep Mohapatra, "How Reliable Are Hog Futures as Forecasts?," *American Journal of Agricultural Economics* 90, no. 2 (2008): 367–78, https://doi.org/10.1111/j.1467-8276.2007.01122.x.

Part III.

Water in a Warming World

A changing climate means that humanity's relationship to water is also changing. Researchers are studying the myriad ways that water and climate change are linked, whether it's higher chances of inland drought or more frequent flooding along the coasts. We are learning that the impacts of climate change are many and varied. In calculating a water budget for the Great Lakes, contributors Gronewold and Rood are finding, for example, that a rapid transition between extreme high and low water levels in the Great Lakes is becoming the "new normal." In their study of ocean wave conditions, Mortlock, Odériz, Mori, and Silva are finding that stronger waves, as a result of climate change, are threatening to erode coastlines. Other researchers are uncovering more holistic pictures of water availability and drought by moving beyond traditional research on river flow to study terrestrial water storage in all forms, as Pokhrel and Felfelani do.

Trees seem to have it especially hard in the United States today, according to the chapters of part III. Scientists Johnson and Partelli-Feltrin are finding that trees under drought stress in the western United States are often too weak to recover from fire damage and are more susceptible to bark beetle infestations. On the East Coast, rising seas are bringing saltwater into freshwater wetlands, where it is killing trees, as Ury describes. Because wetland forests store large quantities of carbon, these forest die-offs will lead to further climate change.

As we devise solutions to adapt to our altered environment, there is much to be learned from strategies under way in cities around the world. Some cities are restricting development on floodplains to minimize human harm and property damage and to create space for rivers to spill over in flood season. Others are embracing managed retreat, or the purposeful movement of houses and businesses away from flood-prone places. Such an approach is not without concern, however, since high-risk areas are often in low-income neighborhoods and communities of color. As Siders and Mach point out in their chapter, managed retreat presents an opportunity to address historical disinvestment and racism in these areas.

As communities embark on bold new experiments to adapt to

a changing climate, researchers caution us to consider unintended consequences of current technologies. For example, we have learned about the limitations of some well-established technologies like levees, as scientists and engineers observe how they can exacerbate floods by pushing river waters to new heights, as Mallea tells us. One highly touted adaptation strategy—building a seawall—proposed for many coastal cities, in response to hurricane surges and rising sea levels, raises important questions, says Griggs, about who will be protected, for how long, and, importantly, who will pay the cost to build them. As an alternative, some researchers are developing green engineering solutions to protect coastlines by mimicking natural breaks and barriers. Others, such as Rhode-Barbarigos and Haus and also Bales, emphasize the need for upgrading natural infrastructure like headwater forests and floodplains. Others still, such as Hare, offer inexpensive citizen-science methods for communities to use—for instance, measuring stream temperature history—in assessing their climate change risk.

The chapters of part III provide a road map of water-related issues brought on by a changing climate and invite us to consider the complex trade-offs at stake as we seek to adapt.

Chapter 12

Two-Thirds of Earth's Land Is on Pace to Lose Water as the Climate Warms— That's a Problem for People, Crops, and Forests

YADU POKHREL and FARSHID FELFELANI

THE WORLD WATCHED with a sense of dread in 2018 as Cape Town, South Africa, counted down the days until the city would run out of water. The region's surface reservoirs were going dry amid its worst drought on record, and the public countdown was a plea for help. By drastically cutting their water use, Cape Town residents and farmers were able to push back "Day Zero" until the rain came, but the close call showed just how precarious water security can be. California

also faced severe water restrictions during its recent multi-year drought.[1] And Mexico City has faced water restrictions after a long dry spell.

There are growing concerns that many regions of the world will face water crises like these in the coming decades as rising temperatures exacerbate drought conditions. Understanding the risks ahead requires looking at the entire landscape of terrestrial water storage—not just rivers but also the water stored in soils, groundwater, snowpack, forest canopies, wetlands, lakes, and reservoirs.

We study changes in the terrestrial water cycle as engineers and hydrologists. In a study published in the journal *Nature Climate Change* in 2021, we and a team of colleagues from universities and institutes around the world showed for the first time how climate change will likely affect water availability on land from all water storage sources over the course of this century.[2] We found that terrestrial water storage is on pace to decline across two-thirds of the land on the planet. The worst impacts will be in areas of the Southern Hemisphere where water scarcity is already threatening food

Cape Town residents queued up for water
as the taps nearly ran dry in 2018.
Morgana Wingard/Getty Images

security and leading to human migration and conflict. Globally, 1 in 12 people could face extreme drought related to water storage every year by the end of this century, compared with an average of about 1 in 33 at the end of the 20th century. These findings have implications for water availability, not only for human consumption, but also for plants, ecosystems, and the sustainability of agriculture.

Where the Risks Are Highest

The water that keeps land healthy, crops growing, and human needs met comes from a variety of sources. Mountain snow and rainfall feed streams that affect community water supplies. Soil water content directly affects plant growth. Groundwater resources are crucial for both drinking water supplies and crop productivity in irrigated regions.

While studies often focus just on river flow as an indicator of water availability and drought, our study instead provides a holistic picture of the changes in total water available on land. That allows us to capture nuances, such as the ability of forests to draw water from deep groundwater sources during droughts when the upper soil levels are drier. The declines we found in land water storage are especially alarming in the Amazon River basin, Australia, Southern Africa, the Mediterranean region, and parts of the United States. In these regions, precipitation is expected to decline sharply with climate change, and rising temperatures will increase evaporation. At the same time, some other regions will become wetter, a process already seen today.[3]

Our findings for the Amazon basin add to the long-standing debate over the fate of the rainforest in a warmer world. Many studies using climate model projections have warned of widespread forest die-off in the future as less rainfall and warmer temperatures lead to higher heat and moisture stress combined with forest fires.[4] In an earlier study, we found that the deep-rooted rainforests may be more resilient to short-term drought than they appear because they can tap water stored in soils deeper in the ground that aren't considered in typical climate model projections.[5] However, findings from 2021, using multiple models, indicate that the declines in total water storage, including deep groundwater stores, may lead to more water shortages during dry seasons when trees need stored water the most and exacerbate future droughts. All this weakens the resilience of the rainforests.

A New Way of Looking at Drought

Our study also provides a new perspective on future droughts.

There are different kinds of droughts. Meteorological droughts are caused by a lack of precipitation. Agricultural droughts are caused by a lack of water in soils. Hydrological droughts involve a lack of water in rivers and groundwater. We provided a new perspective on droughts by looking at the total water storage.

We found that moderate to severe droughts involving water storage would increase until the middle of the 21st century and then remain stable under future scenarios in which countries cut their emissions, but extreme to exceptional water storage droughts could continue to increase until the end of the century. That would further threaten water availability in regions where water storage is projected to decline.

Changes Driven by Global Warming

These declines in water storage and increases in future droughts are primarily driven by climate change, not by land-water management activities such as irrigation and groundwater pumping. This became clear when we examined simulations of what the future would look like if climate conditions were unchanged from preindustrial times. Without the increase in greenhouse gas emissions, terrestrial water storage would remain generally stable in most regions. If future increases in groundwater use for irrigation and other needs are also considered, the projected reduction in water storage and increase in drought could be even more severe.

Notes

1. D. Griffin and K. J. Anchukaitis, "How Unusual Is the 2012–2014 California Drought?," *Geophysical Research Letters* 41, no. 24 (2014): 9017–23, https://doi.org/10.1002/2014gl062433.
2. Y. Pokhrel, F. Felfelani, Y. Satoh, J. Boulange, P. Burek, A. Gädeke, D. Gerten, et al., "Global Terrestrial Water Storage and Drought Severity under Climate Change," *Nature Climate Change* 11, no. 3 (2021): 226–33, https://doi.org/10.1038/s41558-020-00972-w.
3. Nikolaos Skliris, Jan D. Zika, George Nurser, Simon A. Josey, and Robert Marsh, "Global Water Cycle Amplifying at Less Than the Clausius-Clapeyron Rate," *Scientific Reports* 6, no. 1 (2016): https://doi.org/10.1038/srep38752.
4. Yadvinder Malhi, Luiz E. O. C. Aragão, David Galbraith, Chris Huntingford, Rosie Fisher, Przemyslaw Zelazowski, Stephen Sitch, Carol McSweeney, and Patrick Meir, "Exploring the Likelihood and Mechanism of a Climate-Change–Induced Dieback of the Amazon Rainforest," *Proceedings of the National Academy of Sciences*, 106, no. 49 (2009): 20610–15, https://doi.org/10.1073/pnas.0804619106.
5. Yadu N. Pokhrel, Ying Fan, and Gonzalo Miguez-Macho, "Potential Hydrologic Changes in the Amazon by the End of the 21st Century and the Groundwater Buffer," *Environmental Research Letters* 9, no. 8 (2014): 084004, https://doi.org/10.1088/1748-9326/9/8/084004.

Chapter 13

Climate Change Is Making Ocean Waves More Powerful, Threatening to Erode Many Coastlines

THOMAS MORTLOCK, ITXASO ODÉRIZ,

NOBUHITO MORI, and RODOLFO SILVA

SEA LEVEL RISE isn't the only way climate change will devastate the coast. Our research found that climate change is also making waves more powerful, particularly in the Southern Hemisphere.[1] We plotted the trajectory of these stronger waves and found that the coasts of South Australia and Western Australia, Pacific and Caribbean islands, East Indonesia and Japan, and South Africa are already experiencing more powerful waves because of global warming. This will compound the effects of sea level rise by putting low-lying island nations in the Pacific—such as Tuvalu, Kiribati, and the Marshall Islands—in greater danger and by changing how we manage coasts worldwide.

But it's not too late to stop the worst effects—that is, if we drastically and urgently cut greenhouse gas emissions.

An Energetic Ocean

Since the 1970s, the ocean has absorbed more than 90 percent of the heat gained by the planet.[2] This has a range of impacts, including longer and more frequent marine heat waves, coral bleaching, and the provision of energy for more powerful storms.[3]

But our focus was on how warming oceans boost wave power. We looked at wave conditions over the past 35 years and found that global wave power has increased since at least the 1980s, mostly concentrated in the Southern Hemisphere, as more energy is being pumped into the oceans in the form of heat. And a more energetic ocean means larger wave heights and more erosive energy potential for coastlines in some parts of the world than had existed before.

Ocean waves have shaped Earth's coastlines for millions of years. So any small, sustained changes in waves can have long-term consequences for coastal ecosystems and the people who rely on them.[4] Mangroves and salt marshes, for example, are particularly vulnerable to increases in wave energy when combined with sea level rise. To escape, mangroves and marshes naturally migrate to higher ground. But when these ecosystems back onto urban areas, they have nowhere to go and die out. This process is known as "coastal squeeze."

These ecosystems often provide a natural buffer to wave attack for low-lying coastal areas. So without these fringing ecosystems, the coastal communities behind them will be exposed to more wave energy and, potentially, higher erosion.

So Why Is This Happening?

Ocean waves are generated by winds blowing along the ocean surface. And when the ocean absorbs heat, the sea surface warms, encouraging the warm air over the top of it to rise (a movement called convection). This helps spin up atmospheric circulation and winds. In other words, we come to a cascade of impacts: warmer sea surface temperatures bring about stronger winds, which alter global ocean wave conditions.

Our research shows that, in some parts of the world's oceans, wave power is increasing because of stronger wind energy and the shift of westerly winds toward the poles. This is most noticeable in the tropical regions of the Atlantic and Pacific Oceans and in the subtropical regions of the Indian Ocean.

But not all changes in wave conditions are driven by ocean warming from human-caused climate change. Some areas of the world's oceans are still more influenced by natural climate variability—such as by El Niño and La Niña—than by long-term ocean warming.

In general, it appears that changes to wave conditions toward the equator are more driven by ocean warming from human-caused climate change, whereas changes to waves toward the poles remain more impacted by natural climate variability.

How This Could Erode the Coasts

While the response of coastlines to climate change is a complex interplay of many processes, waves remain the principal driver of change along many of the world's open, sandy coastlines. So how might coastlines respond to getting hit by more powerful waves? It generally depends on how much sand there is and how, exactly, wave power increases. If there's an increase in wave height, for example, this may cause increased erosion. But if the waves become longer (a lengthening of the wave period), then this may have the opposite effect, by transporting sand from deeper water to help the coast keep pace with sea level rise.

For low-lying nations in areas of warming sea surface temperatures around the equator, higher waves, combined with sea level rise, pose an existential problem. People in these nations may experience both sea level rise and increasing wave power on their coastlines, eroding land farther up the beach and damaging property. These areas should be regarded as coastal climate hot spots where continued adaption or mitigation funding is needed.

It's Not Too Late

It's not surprising for us to find the fingerprints of greenhouse warming in ocean waves and, consequentially, along our coastlines. Our study looked only at historical wave conditions and how these are already being affected by climate change. But if warming continues in line with current trends over the coming century, we can expect to see more significant changes in wave conditions along the world's coasts than was uncovered in our backward-looking research.

However, if we can mitigate greenhouse warming in line with the 2 degrees Celsius mark of the Paris Agreement, studies indicate we could still keep changes in wave patterns within the bounds of natural climate variability.[5]

Still, one thing is abundantly clear: the impacts of climate change on waves are not a thing of the future, as they are already occurring in large parts of the world's oceans. The extent to which these changes continue and the risk this poses to global coastlines will be closely linked to decarbonization efforts over the coming decades.

Notes

1. I. Odériz, R. Silva, T. R., Mortlock, N. Mori, T. Shimura, A. Webb, R. Padilla-Hernández, and S. Villers, "Natural Variability and Warming Signals in Global Ocean Wave Climates," *Geophysical Research Letters* 48, no. 11 (2021): https://doi.org/10.1029/2021gl093622.

2. S. Levitus, J. Antonov, and T. Boyer, "Warming of the World Ocean, 1955–2003," *Geophysical Research Letters* 32, no. 2 (2005): L02604, https://doi.org/10.1029/2004gl021592.

3. Kevin E. Trenberth, Lijing Cheng, Peter Jacobs, Yongxin Zhang, and John Fasullo, "Hurricane Harvey Links to Ocean Heat Content and Climate Change Adaptation," *Earth's Future* 6, no. 5 (2018): 730–44, https://doi.org/10.1029/2018ef000825.

4. Thomas R. Mortlock, Ian D. Goodwin, John K. McAneney, and Kevin Roche, "The June 2016 Australian East Coast Low: Importance of Wave Direction for Coastal Erosion Assessment," *Water* 9, no. 2 (2017): 121, https://doi.org/10.3390/w9020121.

5. J. Morim, M. Hemer, X. L. Wang, N. Cartwright, C. Trenham, A. Semedo, I. Young, et al., "Robustness and Uncertainties in Global Multivariate Wind-Wave Climate Projections," *Nature Climate Change* 9, no. 9 (2019): 711–18, https://doi.org/10.1038/s41558-019-0542-5.

Chapter 14

As Coastal Flooding Worsens, Some Cities Are Retreating from the Water

A. R. SIDERS and KATHARINE MACH

WHEN THE TIDE gets exceptionally high in Charleston, South Carolina, coastal streets start to run with seawater. Some yards become ponds, and residents pull on gumboots. The city also gets a lot of rain. After homes in one low-lying neighborhood flooded three times in four years, the city offered to buy out 32 flood-prone townhomes and turn the land into open space that can be used for managing future floodwater. It's a strategy that cities with marine waterfronts around the United States are contemplating more often as precipitation patterns change and tidal flooding increases with sea level rise.

Cities all along the US coasts have seen high-tide flooding days double since 2000. By midcentury this is projected to be 5–15 times higher still, with coastal cities experiencing 25–75 high-tide flooding days per year, according to the National Oceanic and Atmospheric Administration's 2021 outlook for high-tide flooding. In 2020, low-lying Charleston saw a record-breaking 14 days of high-tide flooding, and parts of the city experienced even more flood days.

In response, the city has considered seawalls and other measures to try to keep tidal and storm flooding out of threatened neighborhoods. It is elevating some homes. And it has started helping residents relocate away from high-risk areas. This last approach is a strategy known as managed retreat: the purposeful movement of people, buildings, and other infrastructure away from highly hazardous places.[1]

Managed retreat is controversial, particularly in the United

States. But it isn't just about moving—it's about adapting to change and building communities that are safer, addressing long-overlooked needs, and incorporating new technologies and thoughtful design for living and working in today's world. In a special issue of the journal *Science* from 2021, we argued that managed retreat is an opportunity to preserve what communities deem essential while redesigning high-risk areas in ways that are better for everyone.[2]

What Managed Retreat Can Look Like

Managed retreat is an umbrella term that describes a whole category of policies and approaches. The challenge is for communities to pick the approach that makes the most sense for their context. In some places, managed retreat could involve turning streets into canals. In others, it might mean purchasing and demolishing flood-prone properties to create open spaces for stormwater parks that absorb heavy rains or retention ponds and pumping stations. In some cases, managed retreat may involve building denser, more affordable housing that's designed to stay cool during increasingly hot summers, while leaving open spaces for recreation or agriculture that can also reduce heat and absorb stormwater when needed.

However it is done, managing retreat well is challenging. It affects numerous people—the residents who relocate, their neighbors who remain, and the communities where they move—and each may be affected differently.

But managed retreat may also provide opportunities. As US Marine Corps General Oliver P. Smith famously said of a retreat he led during the Korean War, "Retreat, hell! We're just advancing in a different direction." Like General Smith's maneuver, retreat from climate change–related hazards is, at its core, about choosing a new direction. Soldiers Grove, Wisconsin, for example, relocated its flood-prone business district in the late 1970s and used the opportunity to heat the new buildings with solar energy, earning the nickname Solar Village. The move reinvigorated the local economy; yet while the project is hailed as a success, some residents still miss the old town. For managed retreat to be a viable strategy, relocation plans must not only help people move to safer ground but also meet their needs.

At its simplest, managed retreat can be a lifeline for families who are tired of the emotional and financial stress of rebuilding after floods or fires but cannot afford to sell their home at a loss or don't want to sell and put another family at risk. But imagining a new direction for a community could also involve a wide range of social issues, including cultural practices, affordable housing, building codes, land use, jobs, transportation, and utilities. Since high-risk areas are often home to low-income communities and to Black, Indigenous, and other communities of color,[3] addressing climate risk in these areas may also require addressing a national legacy of racism, segregation, and disinvestment that has put these communities at risk and left many with few choices in how to address floods, fires, and other hazards.

Talking about Managed Retreat

Even if an individual or community decides not to retreat, thinking critically and talking openly about managed retreat can help people understand why remaining in place is important and what risks they are willing to face in order to stay. The losses involved in moving can be obvious, but there are losses to staying in place, too: physical risk of future hazards, increased emotional and financial stress,[4] potential loss of community if some residents or businesses leave on their own to find safer ground, pain from watching the environment change,[5] and lost opportunities to improve.

If people can articulate why it is important to remain in place, they can make better plans.

Maybe it is important to stay because a building is historic and people want to protect that history. That opens up creative conversations about the ways people have preserved risk-prone historic buildings and sites. And it invites others to document that heritage and educate the community, perhaps through oral histories, video records, or three-dimensional models.[6]

Maybe it is important for owners to stay because the land has been in the family for generations. That could kick-start conversations with the next generation about their goals for the land and the neighborhood, which may include preserving communities but may also include changes.

Perhaps the local economy depends on the beach, which could start a conversation about why moving back from the beach can be the best way to save the beach and its ecosystem, to prevent walls from narrowing it, and to maintain public access without homes on stilts hovering over the tide.

Maybe a person wants to stay due to deep emotional attachment to a community or to a home. In that case, conversations could focus on moving nearby—to a new house that's safer but still part of the community—or physically relocating the house to a safer place. It could also mean finding strategies, such as life estates, that allow people to stay in their home as long as they want but that prevent a new family from moving in and putting their kids at risk.

In short, thinking carefully about what parts of our lives and communities should stay the same opens space to think creatively about what parts should or could change. And that is the power of thinking about retreat: the power to think creatively about a new direction for a climate-changed world.

Notes

1. Katharine J. Mach and A. R. Siders, "Reframing Strategic, Managed Retreat for Transformative Climate Adaptation," *Science*, 372, no. 6548 (2021): 1294–99, https://doi.org/10.1126/science.abh1894.
2. Brad Wible, "Out of Harm's Way," *Science* 372, no. 6548 (2021): 1274–75, https://doi.org/10.1126/science.abi9209.
3. Robert D. Bullard and Beverly Wright, *The Wrong Complexion for Protection: How the Government Response to Disaster Endangers African American Communities* (New York: New York University Press, 2012).
4. Liz Koslov, Alexis Merdjanoff, Elana Sulakshana, and Eric Klinenberg, "When Rebuilding No Longer Means Recovery: The Stress of Staying Put after Hurricane Sandy," *Climatic Change* 165, no. 3–4 (2021): https://doi.org/10.1007/s10584-021-03069-1.
5. G. Albrecht, G.-M. Sartore, L. Connor, N. Higginbotham, S. Freeman, B. Kelly, H. Stain, A. Tonna, and G. Pollard, "Solastalgia: The Distress Caused by Environmental Change," *Australasian Psychiatry* 15, no. Suppl. 1 (2007): https://doi.org/10.1080/10398560701701288.
6. Tom Dawson, Joanna Hambly, Alice Kelley, William Lees, and Sarah Miller, "Coastal Heritage, Global Climate Change, Public Engagement, and Citizen Science," *Proceedings of the National Academy of Sciences* 117, no. 15 (2020): 8280–86, https://doi.org/10.1073/pnas.1912246117.

Chapter 15

Your Favorite Fishing Stream May Be at High Risk from Climate Change—Here's How to Tell

DANIELLE HARE

MANY OF THE STREAMS that people count on for fishing, water, and recreation are getting warmer as global temperatures rise. But they aren't all heating up in the same way. If communities can figure out which streams will warm the most, they can plan for the future. That has been difficult to predict in the past, but a new method involving temperature patterns may make it easier.

People have widely assumed that streams fed by substantial amounts of groundwater are more resistant to climate change than those fed mostly by snowmelt or rain. It turns out that this groundwater buffering effect varies quite a bit.

The depth of the groundwater affects the response of stream temperature to warming, which in turn affects the habitats of fish and other wildlife and plants.

In a study published in the journal *Nature Communications* in March 2021, my colleagues and I described a simple, inexpensive method that allows communities to look at the temperature history of a stream compared with local air temperature to gauge the depth of the groundwater feeding into it and, from there, assess its risk as the climate changes.[1]

Why Temperature Matters

While a few degrees of temperature change may not seem like much, the majority of animals living in streams and rivers cannot regulate their own body temperatures, so they move around in the environment to find suitable habitats. Many have adapted over time to a narrow range of temperatures. For example, when the waters are warm, especially during hot summer months with low water flow, fish like salmon and trout that live in colder waters must seek out colder water or else perish. These ecological effects can have cascading consequences—for wildlife, humans, and local economies.

Most streams flow all the time. During times without rainfall, water in streams mostly comes from below ground. In fact, groundwater is thought to make up an average of 52 percent of surface water flow across the United States. Because groundwater is typically colder than surface water in summer, the groundwater flowing into streams can buffer the overall stream temperature from climate warming. However, deeper groundwater tends to have more stable temperatures than groundwater closer to the surface.

Previous studies have shown that groundwater temperature is tied to the depth that it travels.[2] Shallow groundwater is more readily influenced by climate variability because it's close to the land's surface.[3] It is also more susceptible to drying, which can reduce, or even disconnect, the shallow groundwater flow from the stream.[4]

Our research builds on these observations. We found that streams with shallow groundwater sources are likely to be warming as much as streams fed mostly by snowmelt and rain, and at similar rates.

Figuring Out a Stream's Risk

The main method currently used to evaluate whether streams are fed by groundwater at large scales cannot differentiate between a stream that relies on shallow groundwater and one fed by deep groundwater. That means that plans for how to manage the effects of climate change are likely not accounting for this important difference. Other studies have also shown that changes to the land, such as from wildfires, snowpack changes, and deforestation, influence shallow groundwater temperature more than deep groundwater temperature.

Looking at temperature patterns can provide more information about the risks that streams might face. We found that when the temperature of a stream follows the same warming and cooling pattern as the air temperature, with a time lag of about 10–40 days, that stream is likely being fed by shallow groundwater. Deeper groundwater stays cooler in the summer, and the stream's temperature doesn't fluctuate as much.

We analyzed the water and air temperature at 1,424 sites along streams across the United States and found that approximately 40 percent of the streams were strongly influenced by groundwater. Of those, we found that half were fed predominantly by shallow groundwater, which was much higher than expected. Comparing this method's results against field and modeling data in smaller studies has shown its rigor.[5]

Because this method requires only stream and air temperature data, landowners and local communities can gather the data at little cost, or it may already be available. Once that information is known, they can plan for future changes and take steps to protect the water quality in streams that are most likely to provide long-term stability.

Notes

1. Danielle K. Hare, Ashley M. Helton, Zachary C. Johnson, John W. Lane, and Martin A. Briggs, "Continental-Scale Analysis of Shallow and Deep Groundwater Contributions to Streams," *Nature Communications* 12, no. 1 (2021): https://doi.org/10.1038/s41467-021-21651-0.
2. Mary P. Anderson, "Heat as a Ground Water Tracer," *Ground Water* 43, no. 6 (2005): 951–68, https://doi.org/10.1111/j.1745-6584.2005.00052.x.
3. B. L. Kurylyk, K. T. MacQuarrie, D. Caissie, and J. M. McKenzie, "Shallow Groundwater Thermal Sensitivity to Climate Change and Land Cover Disturbances: Derivation of Analytical Expressions and Implications for Stream Temperature Modeling," *Hydrology and Earth System Sciences* 19, no. 5 (2015): 2469–89, https://doi.org/10.5194/hess-19-2469-2015.
4. John P. Bloomfield, Benjamin P. Marchant, and Andrew A. McKenzie, "Changes in Groundwater Drought Associated with Anthropogenic Warming," *Hydrology and Earth System Sciences* 23, no. 3 (2019): 1393–1408, https://doi.org/10.5194/hess-23-1393-2019.
5. Martin A. Briggs, Zachary C. Johnson, Craig D. Snyder, Nathaniel P. Hitt, Barret L. Kurylyk, Laura Lautz, Dylan J. Irvine, Stephen T. Hurley, and John W. Lanea, "Inferring Watershed Hydraulics and Cold-Water Habitat Persistence Using Multi-year Air and Stream Temperature Signals," *Science of the Total Environment* 636 (2018): 1117–27, https://doi.or/10.1016/j.scitotenv.2018.04.344.

Chapter 16

Trees Are Dying of Thirst in the Western Drought— Here's What's Going On inside Their Veins

DANIEL JOHNSON and RAQUEL PARTELLI-FELTRIN

LIKE HUMANS, TREES NEED WATER to survive on hot, dry days, and they can survive for only short times under extreme heat and dry conditions. During prolonged droughts and extreme heat waves, even native trees that are accustomed to the local climate can start to die.

Central and northern Arizona have been subject to this. A long-running drought and resulting water stress have contributed to the die-off of as many as 30 percent of the junipers there, according to the US Forest Service. In California, over 129 million trees died as a consequence of a severe drought in

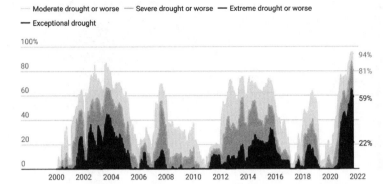

Moderate drought or worse — Severe drought or worse — Extreme drought or worse — Exceptional drought

the last decade, leaving highly flammable dry wood that can fuel future wildfires.[1]

What Happens to Trees during Droughts?

Trees survive by moving water from their roots to their leaves, a process known as vascular water transport. Water moves through small cylindrical conduits, called tracheids or vessels, that are all connected. Drought disrupts the water transport by reducing the amount of water available for the tree. As moisture in the air and soil decline, air bubbles can form in the vascular system of plants, creating embolisms that block the water's flow.

The less water available for trees during dry and hot periods, the higher the chances are that embolisms can form in those water conduits. If a tree can't get water to its leaves, it can't survive.[2] Some species are more resistant to embolisms than others. This is why more piñon pines died in the Southwest during the drought in the early 2000s than juniper trees—junipers are much more resistant.[3]

Drought stress also weakens trees, leaving them sus-

ceptible to bark beetle infestations. During the 2012–2015
drought in the Sierra Nevada, nearly 90 percent of the ponderosa pines died, primarily due to infestations of western
pine beetles.

Drought + Fire Damage Weakens Trees More

Although fire is beneficial for fire-prone forests by controlling
their density and maintaining their health, our research shows
that trees under drought stress are more likely to die from
fires. During droughts, trees have less water for insulation and
cooling against fires. Trees may also reduce their production
of carbohydrates—tree food—during droughts, which leaves
them weaker, making it harder for them to recover from fire
damage.[4]

Trees that suffer trunk damage in a fire are also less likely
to survive in the coming years if drought follows.[5] When trees
have fire scars, their vascular conduits tend to be less functional for water transport around those scars. Traumatic damage to the vascular tissue can also decrease their resistance
to embolisms. So burned trees are more likely to die from
drought, and trees in drought are more likely to die from fire.

What Does This Mean for Future Forests?

Trees in western forests have been dying at an alarming rate
over the past two decades owing to droughts, high tem-

peratures, pests, and fires. As continuing greenhouse gas emissions warm the planet and drive moisture loss, increasing the frequency, duration, and intensity of droughts, research shows that the United States and much of the world will likely witness more widespread tree deaths.

The impact that changing drought and fire regimes will have on forests farther in the future is still somewhat unclear, but several observations may offer some insight.

There is evidence of a transition from forests to shrub-lands or grasslands in parts of the western United States.[6] Frequent burning in the same area can reinforce this transition. When drought or fire alone kills some of the trees, forests often regenerate, but how long it will take for forests to recover to a pre-fire or pre-drought condition after a large-scale die-off or severe fire is unknown.

In the past decade, the western United States has witnessed its most severe droughts in over 1,000 years, including in the Southwest and California. A 2021 study found that subalpine forests in the central Rockies are more prone to fire now than they have been in at least 2,000 years. If there is no change in greenhouse gas emissions, temperatures will continue to increase, and severe drought stress and fire danger days will rise as a result.

Notes

1.   Scott L. Stephens, Brandon M. Collins, Christopher J. Fettig, Mark A. Finney, Chad M. Hoffman, Eric E. Knapp, Malcolm P. North, Hugh Safford, and Rebecca B. Wayman, "Drought, Tree Mortality, and Wildfire in Forests Adapted to Frequent Fire," *BioScience* 68, no. 2 (2018): 77–88, https://doi.org/10.1093/biosci/bix146.
2.   H. D. Adams, M. J. Zeppel, W. R. Anderegg, H. Hartmann, S. M. Landhäusser, D. Tissue, T. E. Huxman, et al., "A Multi-species Synthesis of Physiological Mechanisms in Drought-Induced Tree Mortality," *Nature*

Ecology & Evolution 1, no. 9 (2017): 1285–91, https://doi.org/10.1038/s41559-017-0248-x.

3. Nate McDowell, William T. Pockman, Craig D. Allen, David D. Breshears, Neil Cobb, Thomas Kolb, Jennifer Plaut, et al., "Mechanisms of Plant Survival and Mortality during Drought: Why Do Some Plants Survive While Others Succumb to Drought?," *New Phytologist* 178, no. 4 (2008): 719–39, https://doi.org/10.1111/j.1469-8137.2008.02436.x.

4. Raquel Partelli-Feltrin, Daniel M. Johnson, Aaron M. Sparks, Henry D. Adams, Crystal A. Kolden, Andrew S. Nelson, and Alistair M. S. Smith, "Drought Increases Vulnerability of *Pinus ponderosa* Saplings to Fire-Induced Mortality," *Fire* 3, no. 4 (2020): 56, https://doi.org/10.3390/fire3040056.

5. Raquel Partelli-Feltrin, Alistair M. S. Smith, Henry D. Adams, Crystal A. Kolden, and Daniel M. Johnson, "Short- and Long-Term Effects of Fire on Stem Hydraulics in *Pinus ponderosa* Saplings," *Plant, Cell & Environment* 44, no. 3 (2020): 696–705, https://doi.org/10.1111/pce.13881.

6. Jonathan D. Coop, Sean A. Parks, Camille S. Stevens-Rumann, Shelley D. Crausbay, Philip E. Higuera, Matthew D. Hurteau, Alan Tepley, et al., "Wildfire-Driven Forest Conversion in Western North American Landscapes," *BioScience* 70, no. 8 (2020): 659–73, https://doi.org/10.1093/biosci/biaa061.

Chapter 17

California's Water Supplies Are in Trouble as Climate Change Worsens Natural Dry Spells, Especially in the Sierra Nevada

ROGER BALES

PREPARING FOR YET ANOTHER YEAR of drought is becoming the norm for California, with officials tightening limits on water use to levels normally not seen in past decades. It's becoming common for the state's water reservoirs to begin more years than not well below average, with the outlook for winter rain and snow to replenish supplies being somewhere between uncertain and not promising. While wet years still punctuate multiyear dry periods, they are but partial relief for California's ecosystems and economy, whose year-to-year water demands are much more even.

Especially worrying is the outlook for the Sierra Nevada, the long mountain chain that runs through the eastern part of the state. California's cities and its farms—which grow over a third of the nation's vegetables and two-thirds of its fruit and nuts—rely on runoff from the mountains' snowpack for water.

As an engineer, I have studied California's water and climate for over 40 years. A closer look at California's water resources shows the challenge ahead and how climate change is putting the state's water supply and agriculture at greater risk.

Where California Gets Its Water

Statewide, California averages about 2 feet of precipitation per year, about two-thirds of the global average, giving the state as a whole a semiarid climate. The majority of California's rain and snow falls in the mountains, primarily in winter and spring. But agriculture and coastal cities need that water to get through the dry summers. To get water to dry Southern California and help with flood control in the north, California developed over the past century a statewide system of reservoirs, tunnels, and canals that brings water from the mountains. The largest of those projects, the State Water Project, delivers water from the higher-precipitation northern Sierra to the southern half of the state.

To track where the water goes, it's useful to look at the volume in acre-feet. California is about 100 million acres in area, so at 2 feet per year, its annual precipitation averages about 200 million acre-feet. Of that 200, an average of only about 80 million acre-feet heads downstream. Much of the water returns to the atmosphere through evapotranspiration by plants and trees in the Sierra Nevada or North Coast forests. Of the 80 million acre-feet that does run off, about half remains in the aquatic environment, such as rivers flowing to the ocean. That leaves about 41 million acre-feet for downstream use. About 80 percent of that goes for agriculture and 20 percent for urban uses.

In wet years, there may be much more than 80 million acre-feet of water available, but in dry years, it can be much less. In 2020, for example, California's precipitation was less than two-thirds of average, and the State Water Project delivered only 5 percent of the contracted amounts. The state's other main aqueduct systems that move water around the state also severely reduced their supplies.

The 2021 water year, which ended September 30, was one of the three driest on record for the Sierra Nevada. Precipitation was about 44 percent of average. With limited precipitation as of December 2021 and the state in extreme drought, the State Water Project cut its preliminary allocations for water agencies to 0 percent for 2022, with small amounts still flowing for health and safety needs.

While conditions improve with each winter storm coming in from the Pacific Ocean, the state needs three to five strong storms per year to reach average conditions. The official National Oceanic and Atmospheric Administration seasonal outlook almost always has a one-third chance of a normal year, yet it often assigns more than a one-third likelihood to below-normal precipitation, with above normal being much less likely.

Drought and a Warming Climate

Multiyear dry periods, when annual precipitation is below average, are a feature of California's climate, but rising global temperatures are also having an impact. Over the past 1,100 years, there has been at least one dry period lasting four years or longer each century. There have been two in the past 35 years: 1987–92 and 2012–15. A warmer climate intensifies the effect of these dry periods, as drier soil and drier air stress both natural vegetation and crops.

Rising global temperatures affect runoff from the Sierra Nevada, which provides over 60 percent of California's developed water supply. Over 80 percent of the runoff in the central and southern Sierra Nevada comes from the snow zone. In the wetter but lower-elevation northern Sierra, rainfall contributes over one-third of the annual runoff.[1] The average snowline, the elevation above which most precipitation is snow, goes from about 5,000 feet elevation in the north to 7,000 feet in the south. On average, each 1.8 degrees Fahrenheit (1 Celsius) of warming could push the snowline another 500 feet higher, reducing the snow total.[2]

Shifts from snow to rain and earlier runoff mean that more of the capacity behind existing dams will be allocated to flood control, further reducing their capacity for seasonal water-supply storage. A wealth of research has established that the Sierra Nevada could see low- to no-snow winters for years at a time by the late 2040s if greenhouse gases emissions don't decline, with conditions worsening beyond that being possible.[3] Warming will also increase water demand from forests as growing seasons lengthen.[4] That in turn will drive drought stress leading to tree mortality,[5] as well as an increased risk of high-severity wildfires.

Sustainability in a Warming Climate

Water storage is central to California's water security.

Communities and farms can pump more groundwater

when supplies are low, but the state has been pumping out more water than it replenished in wet years. Parts of the state rely on water from the Colorado River, whose dams provide for several years of water storage, but the basin lacks the runoff to fill the dams. Public opposition has made it difficult to build new dams, so better use of groundwater for both seasonal and multiyear storage is crucial.

The state's Sustainable Groundwater Management Act requires local agencies to develop sustainability plans. That provides some hope that groundwater pumping and replenishment can be brought into balance, most likely by leaving some cropland unplanted. Managed aquifer recharge south of the Sacramento–San Joaquin Delta is gradually expanding, and much more can be done.

If the state doesn't do more, including tactics such as applying desalination technology to make saltwater usable, urban areas can expect the 25 percent cuts in water use put in place during the 2012–15 drought to be more common and potentially even deeper.

California's water resources can provide for a healthy environment, robust economy, and sustainable agricultural use. Achieving this will require upgrading both natural infrastructure—headwaters forests, floodplains, and groundwater recharge in agricultural areas—and built infrastructure, such as canals, spillways, and levees. The information is available; we all now have to follow through.

Notes

1. Joseph Rungee, Qin Ma, Michael Goulen, and Roger Bales, "Evapotranspiration and Runoff Patterns across California's Sierra Nevada," *Frontiers in Water* 3 (2021): https://doi.org/10.3389/frwa.2021.655485.
2. James W. Roche, Roger C. Bales, Robert Rice, and Danny G. Marks, "Management Implications of Snowpack Sensitivity to Temperature and

Atmospheric Moisture Changes in Yosemite National Park, CA," *Journal of the American Water Resources Association* 54, no. 3 (2018): 724–41, https://doi.org/10.1111/1752-1688.12647.

3. Erica R. Siirila-Woodburn, Alan M. Rhoades, Benjamin J. Hatchett, Laurie S. Huning, Julia Szinai, Christina Tague, Peter S. Nico, et al., "A Low-to-No Snow Future and Its Impacts on Water Resources in the Western United States," *Nature Reviews Earth & Environment* 2, no. 11 (2021): 800–819, https://doi.org/10.1038/s43017-021-00219-y.

4. Michael L. Goulden and Roger C. Bales, "Mountain Runoff Vulnerability to Increased Evapotranspiration with Vegetation Expansion," *Proceedings of the National Academy of Sciences* 111, no. 39 (2014): 14071–75, https://doi.org/10.1073/pnas.1319316111.

5. M. L. Goulden and R. C. Bales, "California Forest Die-off Linked to Multi-year Deep Soil Drying in 2012–2015 Drought," *Nature Geoscience* 12, no. 8 (2019): 632–37, https://doi.org/10.1038/s41561-019-0388-5.

Chapter 18

For Flood-Prone Cities, Seawalls Raise as Many Questions as They Answer

GARY GRIGGS

THE OCEANS ARE RISING at an accelerating rate,[1] and millions of people are in the way. Rising tides are already affecting cities along low-lying shorelines, such as the US Atlantic and Gulf coasts, where sunny-day flooding has become common during high tides.

The US Army Corps of Engineers, whose mission includes maintaining waterways and reducing disaster risks, has proposed building large and expensive seawalls to protect a number of US cities, neighborhoods, and shorelines from coastal storms and rising seas. Charleston, New York City, and the Houston-Galveston metro area are currently considering proposals to build barriers in response to hurricane surges and sea level rise, and the Corps in 2021 published a draft proposal of a seawall for Miami.

As a scientist who studies the evolution and development of coastlines and the impacts of sea level rise, I believe that large-scale seawall proposals raise important long-term questions that residents, urban leaders, and elected officials at all levels of government need to consider carefully before they invest billions of dollars. In my view, this approach is almost certainly a short-term strategy that will protect only a few cities and only selected portions of those cities effectively.

Coastal Flooding Is Here

The extent of high-tide flooding in low-elevation Atlantic coastal cities is well documented, and so are future trends. In a 2017 study, the Union of Concerned Scientists assessed chronic flooding risks in 52 large coastal cities and found that, by 2030, the 30 cities most at risk can expect at least two dozen tidal floods yearly on average.[2] The study defined tidal flooding as seawater encroaching into at least 10 percent of a city. These cities include Boston; New Haven, Connecticut; Philadelphia; Wilmington, Delaware; Baltimore; Washington, DC; Norfolk, Virginia; Charleston, South Carolina; Savannah, Georgia; Jacksonville, Florida; and Miami. Cumulatively, they are home to about six million people. The study projected that, by 2045, most of them will experience over 100 days of flooding annually.

This flooding won't just become more frequent—it also will become deeper, extend farther inland, and last longer as sea levels continue to rise. Greater encroachment will cause increasing harm to infrastructure, development, and property.

The Army Corps of Engineers' proposals include building an 8-mile (12.8-kilometer) seawall around Charleston at a cost of nearly US$2 billion; a 1-mile (1.6-kilometer) wall for Miami-Dade, with a price tag of nearly US$4.6 billion; and a 6-mile (9.6-kilometer) barrier to shield portions of New York City and New Jersey, at an estimated cost of US$119 billion.[3] None of these investments would protect other Atlantic coastal cities already experiencing high-tide flooding.[4] While these proposed projects differ slightly, they each involve major barriers or seawalls along the shoreline, or just offshore in the case of New York and New Jersey. The structures are intended to protect these areas from hurricanes and storm surges and from some uncertain level of future sea level rise.

Coverage and Costs

There are a number of key issues that I believe any city considering a major seawall proposal should consider. Here are some of the most critical questions:

- Who and what will be protected by these large walls and at what cost? With so many US cities already experiencing coastal flooding, and current proposals focusing on metropolitan areas, there are important questions about which portions of cities would be surrounded by walls and how much to spend. For example, New York City has a 520-mile coastline, but seawall proposals there focus only on protecting lower Manhattan.
- How many years of protection might these barriers

provide, and are they just short-term solutions? Flooding can result from short-term extreme events, such as hurricanes, and also from long-term sea level rise. What time frame should these projects be designed to address?

- Who selects which cities or areas to protect? To date, proposals have come from the Army Corps of Engineers. Which officials and local, state, or federal agencies should be involved in making these decisions and establishing policies that will guide responses to future sea level rise?
- Do people really want to live behind walls? In New York, Miami, and elsewhere, residents have objected to flood walls that would block views.
- Who will pay for the walls? Proposing multibillion-dollar walls is one thing, but where will the funds come from to construct and maintain these massive structures? In Texas, where the proposed "Ike Dike" across Galveston Bay is projected to cost some US$26 billion, the legislature is considering creating a special flood-control district with the power to levy property taxes within its boundaries in order to raise the state's share.[5]
- Would these structures encourage additional development behind the walls? Typically, providing flood control encourages new construction in the now-protected area, which increases future liabilities and losses when walls are overtopped or fail.
- What other long-term options should be considered? Boston recently considered flood barriers for either its outer or inner harbor but rejected these options in favor of softer options like climate-resilient zoning with

special requirements for new projects in flood-prone areas.

- Can cities that reject seawalls agree on thresholds or trigger points for taking other steps, such as using some combination of incentives and mandates to move people out of high-risk areas? Norfolk, one of the most flood-prone cities on the Atlantic coast, has developed a plan that prioritizes development in less-vulnerable zones, which it calls "neighborhoods of the future."

Such decisions will affect coastal communities, infrastructure, and residents for decades into the future, and I believe it is time to meet this crisis head-on. Sea level rise is a complex problem with no easy or inexpensive solution, but the sooner the science is understood and accepted, and everyone who is affected has an opportunity to get involved, the sooner that cities can make plans. In the long run, there is no way to hold back the Atlantic Ocean.

Notes

1. T. Veng and O. B. Andersen, "Consolidating Sea Level Acceleration Estimates from Satellite Altimetry," *Advances in Space Research* 68, no. 2 (2021): 496–503, https://doi.org/10.1016/j.asr.2020.01.016.
2. "When Rising Seas Hit Home," Union of Concerned Scientists, July 5, 2017, https://www.ucsusa.org/resources/when-rising-seas-hit-home.
3. Anne Barnard, "The $119 Billion Sea Wall That Could Defend New York . . . or Not," *New York Times*, January 17, 2020.
4. William Sweet, Gregory Dusek, Greg Carbin, John Marra, Doug Marcy, and Steven Simon, *2019 State of U.S. High Tide Flooding with a 2020 Outlook* (Silver Spring, MD: National Oceanic and Atmospheric Administration, July 2020), https://tidesandcurrents.noaa.gov/publications/Techrpt_092_2019_State_of_US_High_Tide_Flooding_with_a_2020_Outlook_30June2020.pdf.
5. Nick Powell and Taylor Goldenstein, " 'Ike Dike' Funding Bill Nears Approval in Texas Legislature," *Houston Chronicle*, May 23, 2021, https://www.houstonchronicle.com/news/houston-texas/houston/article/Bill-to-create-taxing-entity-to-pay-for-Ike-16197375.php.

Chapter 19

A 20-Foot Seawall Won't Save Miami— How Living Structures Can Help Protect the Coast and Keep the Paradise Vibe

LANDOLF RHODE-BARBARIGOS and BRIAN HAUS

MIAMI IS ALL ABOUT THE WATER and living life outdoors. Walking paths and parks line large stretches of downtown waterfront with a stunning bay view. This downtown core is where the US Army Corps of Engineers developed plans to build a US$6 billion seawall, 20 feet high in places, through downtown neighborhoods and right between the Brickell district's high-rises and the bay. There's no question that the city is at increasing risk of flooding as sea level rises and storms intensify with climate change. A hurricane as powerful as 1992's Andrew or 2017's Irma making a direct hit on Miami would devastate the city.[1]

But the seawall the Army Corps included in its original proposal from 2021—protecting only 6 miles of downtown and the financial district from a storm surge—would not save Miami-Dade County. Most of the county would have been outside the wall, unprotected; the wall would trap water inside the city; and the Corps did not closely study what the construction of a high sea wall would do to water quality. At the same time, it would block the water views that the city's economy thrives on.

To protect more of the region without losing Miami's vibrant character, there are ways to pair the strength of less obtrusive hard infrastructure with nature-based "green" solutions. With our colleagues at the University of Miami's Climate Resilience Academy, including those from the Rosenstiel School of Marine and Atmospheric Science and the College of Engineering, we have been designing and testing innovative hybrid solutions.

Much of Miami is built right up to the water's edge.
On average, it's 6 feet above sea level.
Ryan Parker / Unsplash, CC BY-ND

Natural Storm Management

Living with water today doesn't look the same as it did 50 years ago, or even 20 years ago. Parts of Miami now regularly see "sunny day" flooding during high tides. Saltwater infiltrates basements and high-rise parking garages, and tidal flooding is forecast to occur more frequently as sea level rises. When storms come through, the storm surge adds to that already high water. Hurricanes are less common than tidal flooding, but their destructive potential is greater, and that is what the Army Corps is focused on with its seawall plan. If Miami Beach were an undeveloped barrier island, and if thick mangrove forests were still common along the South Florida shoreline, the Miami area would have more natural protection against storm surge and wave action. But living buffers are mostly long gone.

There are still ways that nature can help preserve the beauty of Miami's marine playground, though.

For example, healthy coral reefs break incoming waves, dissipating their energy before the waves reach shore. Dense mangrove forests also dissipate wave energy with their complex root systems that rise above the water line, dramatically reducing the waves' impact.[2] In areas where coastal flooding is an increasing problem, low-lying communities can be relocated to higher ground and the vacant land be turned into wetlands, canals, or parks that are designed to manage flooding from storm surges.

Each area of coastline is unique and requires different

protective measures based on the dynamics of how the water flows in and out, the local ecosystems, and the built environment one aims to protect. Given Miami's limited space, living shorelines alone won't be enough protection against a major hurricane, but there are powerful ways to pair them with solid "gray" infrastructure that are more successful than either alone.

Hybrid Solutions Mix Green and Gray

Nobody wants to look at a concrete breakwater structure offshore. But if you're offered a breakwater below the surface, covered with corals and hospitable to other marine life, a place where you can snorkel, that's different. Corals help the structure dissipate wave energy better,[3] and at the same time, they improve water quality, habitat, recreation, tourism, and quality of life. For a lot of people, those are some of Miami's main selling points. By pairing corals and mangroves with a more sustainable and eco-friendly hard infrastructure, hybrid solutions can be far less obtrusive than a tall seawall.

For example, a cement-based breakwater structure submerged offshore with coral transplants could provide habitat for entire ecosystems while also providing protection for the shore.[4] We've worked with the City of Miami Beach through the University of Miami Laboratory for Integrative Knowledge to implement three hybrid coral reefs that we will monitor for their engineering and ecological performance.

Closer to shore, we're experimenting with a novel modular marine and estuarine system we call "SEAHIVE." Below the water line, water flows through hollow hexagonal channels of concrete, losing energy. The top can be filled with soil to grow coastal vegetation such as mangroves, thereby providing even more protection as well as an ecosystem that benefits the bay. SEAHIVE is a green engineering alternative under research and development for riprap in North Bay Village, an inhabited island in the bay, and serves as infrastructure in a newly developed marine park in Pompano Beach called Wahoo Bay, developed in collaboration with local governments and Shipwreck Park, a not-for-profit organization.

What about the Rest of Miami?

In September 2021, Miami-Dade County rejected the Army Corps of Engineers' draft plan, opting instead to devise its own plan.

The original plan would have given nature-based solutions little role beyond a fairly small mangrove and seagrass restoration project to the south. The Corps determined that natural solutions alone would require too much space and wouldn't be as effective as hard infrastructure in a worst-case scenario. So, the Army Corps' plan focused on the 6-mile seawall, floodgates, and elevating or strengthening buildings. The draft plan proposed basically to protect the downtown infrastructure and leave everyone else on their own.

Permanent seawalls and floodgates can also affect water flow and harm water quality. The Corps' own documents warn that the seawalls and gates would affect wildlife and ecosystems, including loss of protective corals, mangroves, and seagrass beds. We would like to see a plan for all of Miami-Dade County that explores the protective services of green and hybrid (green-gray) solutions and also considers their value for marine life, tourism, fishing, and general quality of life.

Both types—green and gray—would take time to build out. And both run a risk of failure from global and local issues, including climate change and diseases. Corals can die from high ocean temperatures, ocean acidification, and pollution, while sea level rise and storms can damage mangroves; but storms can also undermine engineered solutions, like the New Orleans levee system during Hurricane Katrina. To help build resilience, our colleagues at the University of Miami have been breeding corals to be more resistant to climate change, investigating novel cementitious materials and noncorrosive reinforcements, and developing new designs for coastal structures.[5]

Miami in the Future

Miami will be different in the coming decades, and the changes are already starting.

High ground is at a premium, and that's showing up in real estate decisions that are pushing lower-income residents out and into less safe areas. Anybody looking back at Miami's history will probably think the region should have done a better job of managing growth and maybe even managing some form of retreat from threatened areas.

We don't want to see Miami become a Venice or a city walled off from the water. We think Miami can thrive by making use of the local ecosystem with novel green engineering solutions and an architecture that adapts.

Notes

1. Jessica Weinkle, Chris Landsea, Douglas Collins, Rade Musulin, Ryan P. Crompton, Philip J. Klotzbach, and Roger Pielke Jr., "Normalized Hurricane Damage in the Continental United States, 1900–2017," *Nature Sustainability* 1, no. 12 (2018): 808–13, https://doi.org/10.1038/s41893 -018-0165-2.
2. Ahmad Mustafa Hashim and Sim Mong Pheng Catherine, "A Laboratory Study on Wave Reduction by Mangrove Forests," *APCBEE Procedia* 5 (2013): 27–32, https://doi.org/10.1016/j.apcbee.2013.05.006.

3. Mohammad Ghiasian, Jane Carrick, Claire Bisson, Brian K. Haus, Andrew C. Baker, Diego Lirman, and Landolf Rhode-Barbarigos, "Laboratory Quantification of the Relative Contribution of Staghorn Coral Skeletons to the Total Wave-Energy Dissipation Provided by an Artificial Coral Reef," *Journal of Marine Science and Engineering* 9, no. 9 (2021): 1007, https://doi.org/10.3390/jmse9091007.

4. Mohammad Ghiasian, Jane Carrick, Landolf Rhode-Barbarigos, Brian Haus, Andrew C. Baker, and Diego Lirman, "Dissipation of Wave Energy by a Hybrid Artificial Reef in a Wave Simulator: Implications for Coastal Resilience and Shoreline Protection," *Limnology and Oceanography: Methods* 19, no. 1 (2020): 1–7, https://doi.org/10.1002/lom3.10400.

5. Kathleen Sullivan Sealey, Esber Andiroglu, Joel Lamere, James Sobczak, and Prannoy Suraneni, "Multifunctional Performance of Coastal Structures Based on South Florida Coastal Environs," *Journal of Coastal Research* 37, no. 3 (2021): https://doi.org/10.2112/jcoastres-d-20-00068.1.

Chapter 20

Sea Level Rise Is Killing Trees along the Atlantic Coast, Creating "Ghost Forests" That Are Visible from Space

EMILY URY

TREKKING OUT TO MY RESEARCH SITES near North Carolina's Alligator River National Wildlife Refuge, I slog through knee-deep water on a section of trail that is completely submerged. Permanent flooding has become commonplace on this low-lying peninsula, nestled behind North Carolina's Outer Banks. The trees growing in the water are small and stunted. Many are dead.

Throughout coastal North Carolina, evidence of forest

die-off is everywhere. Nearly every roadside ditch I pass while driving around the region is lined with dead or dying trees. As an ecologist studying wetland response to sea level rise, I know this flooding is evidence that climate change is altering landscapes along the Atlantic coast. It's emblematic of environmental changes that also threaten wildlife, ecosystems, and local farms and forestry businesses.

Like all living organisms, trees die. But what is happening here is not normal. Large patches of trees are dying simultaneously, and saplings aren't growing to take their place. And it's not just a local issue: Seawater is raising salt levels in coastal woodlands along the entire Atlantic Coastal Plain, from Maine to Florida. Huge swaths of contiguous forest are dying. They're now known in the scientific community as "ghost forests."

The Insidious Role of Salt

Sea level rise driven by climate change is making wetlands wetter in many parts of the world. It's also making them saltier.[1]

In 2016 I began working in a forested North Carolina wetland to study the effect of salt on its plants and soils. Every couple of months, I suit up in heavy rubber waders and a mesh shirt for protection from biting insects, and I haul over 100 pounds of salt and other equipment out along the flooded trail to my research site. We are salting an area about the size of a tennis court, seeking to mimic the effects of sea level rise.

After two years of effort, the salt didn't seem to be affecting the plants or soil processes that we were monitoring. I realized that instead of waiting around for our experimental salt to slowly kill these trees, the questions I needed to answer were how many trees had already died and how much more wetland area was vulnerable. To find answers, I had to go to sites where the trees were already dead.

Rising seas are inundating North Carolina's coast, and saltwater is seeping into wetland soils. Salts move through groundwater during phases when freshwater is depleted, such as during droughts. Saltwater also moves through canals and ditches, penetrating inland with help from wind and high tides. Dead trees with pale trunks, devoid of leaves and limbs, are a telltale sign of high salt levels in the soil. A 2019 report called them "wooden tombstones."

As the trees die, more salt-tolerant shrubs and grasses move in to take their place. In a 2021 study that I coauthored with Emily Bernhardt and Justin Wright at Duke University and Xi Yang at the University of Virginia, we show that in North Carolina this shift has been dramatic.[2] The state's coastal region has suffered a rapid and widespread loss of forest, with cascading impacts on wildlife, including the endangered red wolf and red-cockaded woodpecker. Wetland forests sequester and store large quantities of carbon,[3] so forest die-offs also contribute to further climate change.

Assessing Ghost Forests from Space

To understand where and how quickly these forests are changing, I needed a bird's-eye perspective. This perspective comes from satellites like NASA's Earth Observing System, which are important sources of scientific and environmental data.

Since 1972, Landsat satellites, jointly operated by NASA and the US Geological Survey, have captured continuous images of Earth's land surface that reveal both natural and human-induced change. We used Landsat images to quantify changes in coastal vegetation since 1984 and referenced high-resolution Google Earth images to spot ghost forests. Computer analysis helped identify similar patches of dead trees across the entire landscape.

The results were shocking. We found that more than 10 percent of forested wetland within the Alligator River National Wildlife Refuge was lost over the past 35 years. This is federally protected land, with no other human activity that could be killing off the forest. Rapid sea level rise seems to be outpacing the ability of these forests to adapt to wetter, saltier conditions. Extreme weather events, fueled by climate change, are causing further damage from heavy storms, more frequent hurricanes, and drought.

We found that the largest annual loss of forest cover within our study area occurred in 2012, following a period of extreme drought, forest fires, and storm surges from Hurricane Irene in August 2011. This triple whammy seemed to have been a tipping point that caused mass tree die-offs across the region.

Should Scientists Fight the Transition or Assist It?

As global sea levels continue to rise, coastal woodlands from the Gulf of Mexico to the Chesapeake Bay and elsewhere around the world could also suffer major losses from saltwater intrusion.[4] Many people in the conservation community are rethinking land management approaches and exploring more adaptive strategies, such as facilitating forests' inevitable transition into salt marshes or other coastal landscapes.[5]

For example, in North Carolina the Nature Conservancy is carrying out some adaptive management approaches, such as creating "living shorelines" made from plants, sand, and rock to provide natural buffering from storm surges. A more radical approach would be to introduce marsh plants that are salt-tolerant in threatened zones. This strategy is controversial because it goes against the desire to try to preserve ecosystems exactly as they are.

But if forests are dying anyway, having a salt marsh is a far better outcome than allowing a wetland to be reduced to open water. While open water isn't inherently bad, it does not provide the many ecological benefits that a salt marsh affords. Proactive management may prolong the life span of coastal wetlands, enabling them to continue storing carbon, providing habitat, enhancing water quality, and protecting productive farm and forest land in coastal regions.

Notes

1. Michael Oppenheimer, Bruce C. Glavovic, Jochen Hinkel, Roderik van de Wal, Alexandre K. Magnan, Amro Abd-Elgawad, Rongshuo Cai, et al., "Sea Level Rise and Implications for Low-Lying Islands, Coasts and Communities," chap. 4 in *IPCC Special Report on the Ocean and Cryosphere in a Changing Climate* (Cambridge: Cambridge University Press, 2019), 321–445, https://doi.org/10.1017/9781009157964.006.
2. Emily A. Ury, Xi Yang, Justin P. Wright, and Emily S. Bernhardt, "Rapid Deforestation of a Coastal Landscape Driven by Sea-Level Rise and Extreme Events," *Ecological Applications* 31, no. 5 (2021): https://doi.org/10.1002/eap.2339.
3. Lindsey S. Smart, Paul J. Taillie, Benjamin Poulter, Jelena Vukomanovic, Kunwar K. Singh, Jennifer J. Swenson, Helena Mitasova, Jordan W. Smith, and Ross K. Meentemeyer, "Aboveground Carbon Loss Associated with the Spread of Ghost Forests as Sea Levels Rise," *Environmental Research Letters* 15, no. 10 (2020): 104028, https://doi.org/10.1088/1748-9326/aba136.
4. Elliott White and David Kaplan, "Restore or Retreat? Saltwater Intrusion and Water Management in Coastal Wetlands," *Ecosystem Health and Sustainability* 3, no. 1 (2017): https://doi.org/10.1002/ehs2.1258.
5. Sam Morgan, "Adaptive Planning for Coastal Climate Change," Landscape Architecture Aotearoa, September 10, 2018, https://www.landscapearchitecture.nz/landscape-architecture-aotearoa/2018/9/7/adaptive-planning-for-coastal-climate-change.

Chapter 21

Climate Change Is Driving Rapid Shifts between High and Low Water Levels on the Great Lakes

DREW GRONEWOLD and RICHARD B. ROOD

THE NORTH AMERICAN GREAT LAKES contain about one-fifth of the world's surface freshwater. In May 2019, new records for high water level were set on Lakes Erie and Superior, with widespread flooding across Lake Ontario for the second time in three years.[1] These events coincided with persistent precipitation and severe flooding across much of central North America.

As recently as 2013, water levels on most of the Great Lakes were very low. At that time some experts proposed that climate change, along with other human actions such as channel dredging and water diversions, would cause water levels to continue to decline.[2] This scenario spurred serious concern. Over 30 million people live within the Great Lakes basin, and many depend directly on the lakes for drinking water, industrial use, commercial shipping, and recreation.

High water poses just as many challenges as too little water for the region, including shoreline erosion, property damage, displacement of families, and delays in planting spring crops. In May 2019, New York governor Andrew Cuomo declared a state of emergency in response to the flooding around Lake Ontario while calling for better planning decisions in light of climate change.

As researchers specializing in hydrology and climate science, we believe rapid transitions between extreme high and low water levels in the Great Lakes are a consequence of the warming climate. Our view is based on interactions between global climate variability and the components of the regional hydrological cycle. Increasing precipitation, the threat of recurring periods of high evaporation, and a combination of both routine and extreme weather events—such as extreme cold air outbursts—are putting the region in uncharted territory.

Calculating the Lakes' Water Budget

In 2019, water levels on the Great Lakes were setting records. Lake Superior, the largest freshwater lake on Earth by surface area, surpassed its record of 602.82 feet for the month of May and would set a new record for the month of June. Lake

Erie, the world's ninth largest lake by surface area, surpassed not only its record water level for the month of May but also its all-time monthly water level record of 574.28 feet, which had stood since June 1986. This record was, subsequently, broken several times in 2019 and 2020.

These extremes result from changes in the Great Lakes' water budget: the movement of water into and out of the lakes. Water levels across the lakes fluctuate over time, influenced mainly by three factors, namely, rain and snowfall over the lakes, evaporation over the lakes, and runoff that enters each lake from the surrounding land through tributaries and rivers. Runoff is directly affected by precipitation over land, snow cover, and soil moisture.

Interactions among these factors drive changes in the amount of water stored in each of the Great Lakes. For example, in the late 1990s, surface water temperatures on Lakes

Superior and Michigan–Huron rose by roughly 2 degrees Celsius. Water evaporates more rapidly when it is warmer, and during this period, evaporation rates were nearly 30 percent above annual average levels. Water levels on Lake Michigan–Huron dropped to the lowest levels ever recorded.[3]

Then in 2014 the Midwest experienced an extraordinary cold air outbreak, widely dubbed the "polar vortex." The lakes froze and evaporation rates dropped. As a result, water levels surged.

At roughly the same time, precipitation was increasing. The 2017 Lake Ontario flood followed a spring of extreme overland precipitation in the Lake Ontario and Saint Lawrence River basins. The 2019 flood followed the wettest US winter in history.

What do these trends mean for water levels? In addition to the current onset of record highs, water levels in Lake Erie have been rising earlier in spring and declining earlier in fall.[4] More winter precipitation is falling, often as snow. The snow is melting earlier in response to rising temperatures and shorter winters. The resulting runoff is then amplified in years like 2019 with large springtime rains. The net effect of this combination of hydrological events is that Lake Erie's current water levels are much higher than usual for this time of year.

The Role of Climate Change

Great Lakes water levels have varied in the past, so how do we know whether climate change is a factor in the changes taking place now?

Precipitation increases in winter and spring are consistent with the fact that a warming atmosphere can transport more water vapor. Converting water from vapor to liquid and ice releases energy. As a result, increased atmospheric moisture contributes to more precipitation during extreme events. That is, when weather patterns are wet, they are very wet.

Changes in seasonal cycles of snowmelt and runoff align with the fact that spring is coming earlier in a changing climate. Climate models project that this trend will continue. Similarly, rising lake temperatures contribute to increased evaporation. When weather patterns are dry, this produces lower lake levels.

Wet and dry periods are influenced by storm tracks, which are related to global-scale processes such as El Niño. Similarly, cold air outbreaks are related to the Arctic Oscillation and associated shifts in the polar jet stream. These global patterns often have indirect effects on Great Lakes weather.[5] It is uncertain how these relationships will change as the planet warms.

Tools for Better Forecasts

Rapid changes in weather and water supply conditions across the Great Lakes and upper Midwest are already challenging water management policy, engineering infrastructure, and human behavior. We are undoubtedly observing the effects of a warming climate in the Great Lakes, but many questions remain to be answered.

The Great Lakes are, collectively, a critical water resource. Government agencies and weather forecasters need new tools to assess how future climate conditions may affect the Great Lakes water budget and water levels, along with better shorter-term forecasts that capture changing conditions. Innovative techniques, such as incorporating information from snow and soil moisture maps into seasonal water supply forecasts, can help capture a full picture of what is happening to the water budget. The bigger point is that past conditions around the Great Lakes are not a reliable basis for decision-making that will carry into the future.

Notes

1. Andrew D. Gronewold and Richard B. Rood, "Recent Water Level Changes across Earth's Largest Lake System and Implications for Future Variability," *Journal of Great Lakes Research* 45, no. 1 (2019): 1–3, https://doi .org/10.1016/j.jglr.2018.10.012.
2. Katharine Hayhoe, Jeff VanDorn, Thomas Croley II, Nicole Schlegal, and Donald Wuebbles, "Regional Climate Change Projections for Chicago and the US Great Lakes," *Journal of Great Lakes Research* 36 (201): 7–21, https://doi.org/10.1016/j.jglr.2010.03.012.
3. Andrew D. Gronewold and Craig A. Stow, "Water Loss from the Great Lakes," *Science* 343, no. 6175 (2014): 1084–85, https://doi.org/10.1126 /science.1249978.
4. Andrew D. Gronewold and Craig A. Stow, "Unprecedented Seasonal Water Level Dynamics on One of the Earth's Largest Lakes," *Bulletin of the American Meteorological Society* 95, no. 1 (2014): 15–17, https://doi .org/10.1175/bams-d-12-00194.1.
5. Elizabeth Carter and Scott Steinschneider, "Hydroclimatological Drivers of Extreme Floods on Lake Ontario," *Water Resources Research* 54, no. 7 (2018): 4461–78, https://doi.org/10.1029/2018wr022908.

Chapter 22

As Flood Risks Increase across the US, It's Time to Recognize the Limits of Levees

AMAHIA MALLEA

MANY CITIES RELY ON LEVEES for flood protection. There are more than 100,000 miles of levees nationwide, in all 50 states and one of every five counties. Most of them seriously need repair: levees got a D on the American Society of Civil Engineers' 2021 national infrastructure report card.

Levees shield towns and farms from flooding, but they also create risk. When rivers rise, they can't naturally spread out on the floodplain as they did in the pre–flood control era. Instead, they flow harder and faster and send more water downstream.

And climate models show that flood risks are increasing. The Midwest will be warmer and wetter, like in 2008 and 2019 when dozens of levees on the Missouri, Mississippi, and Arkansas Rivers were overtopped or breached by floodwaters.[1] Across the central United States, rivers are becoming increasingly hard to control.

Remaking the Missouri

In my 2018 book, *A River in the City of Fountains*, I describe the complexities of flood control in Kansas City, which sits at the junction of the Missouri and Kansas Rivers. The Missouri, the larger of the two, is America's longest river, rising in Montana's Rocky Mountains and flowing east and south for 2,341 miles until it joins the Mississippi River north of St. Louis. Historically it was wide and shallow, full of sand bars and snags that created challenges for steamboats.

In the late 19th and early 20th centuries, Kansas City business leaders began lobbying for federal navigation subsidies to counter the influence of the railroads. Until the river could be narrowed and deepened, navigation was unreliable. And without levees, industry in the floodplain was at risk. Major floods inundated Kansas City in 1903, 1908, 1943, and 1951, leaving thousands homeless and causing heavy economic damage. These disasters convinced civic leaders that more was needed than piecemeal navigation and flood-control projects along the lower Missouri.

In 1944 they got their wish when Congress passed the Flood Control Act, which authorized the construction of dozens of dams nationwide. One section of the bill, the Missouri Basin Plan, sought to convert the entire Missouri into what historian Donald Worster calls an "ornate hydraulic regime," with five upstream dams for hydroelectric power, irrigation, and recreation, as well as levees and a navigable barge channel from Sioux City to St. Louis.[2]

Over the next decade, engineers built levees and straightened and dredged the river channel. Upstream dams curbed the Missouri's spring rise. In August 1955, *Life* magazine reported that "U.S. engineers have finally clinched their victory over the rampaging Missouri River. Flood control . . . is already bringing prosperity to the valley it drains."

The Limits of Levees

Today Kansas City and many other US river towns are fortified behind levees and floodwalls, but faith in the idea of engineered flood control is starting to erode. Disastrous midwestern flooding in the summer of 1993, which killed 50 people and caused US$15 billion in damages, showed the limitations of this strategy. Floodwaters rose to unprecedented levels, eventually breaching or overtopping more than 1,000 levees. After the waters ebbed, federal and state officials paid to move some homes and communities off floodplains to higher ground. However, this trend quickly reversed. By 2008, Missouri had authorized more than $2 billion of new development in zones that were flooded in 1993.

Many Kansas City residents still believe that higher, stronger levees will hold back future floods, and Congress has authorized millions of dollars to build them. But experienced engineers like retired Brigadier General Gerald Galloway of the US Army, who coauthored a federal government assessment of the 1993 floods, warn that "there's no such thing as absolute protection."[3]

For their part, many scientists and engineers have found

that levees can exacerbate floods by pushing river waters to new heights. One 2018 study estimated that about 75 percent of increases in the magnitude of 100-year floods on the lower Mississippi River over the past 500 years could be attributed to river engineering.[4]

What about commercial benefits from channeling rivers? Kansas City is still an economic hub, but railroads and highways have been more important than barges. The Missouri carries only a fraction of the tonnage shipped on other navigable rivers, such as the Mississippi, even though its channel has been expensively built and maintained for over 100 years.

Rethinking River Control

Levees also constrain cities' relationships with rivers, walling off any connections for purposes other than commerce. Author William Least Heat-Moon captured this paradox when he traveled across the United States by boat in the late 1990s and observed that "Kansas City, born of the Missouri, has turned away from its great genetrix more than almost any other river city in America."[5] More recently, however, Kansas City has begun to remember its interest and love for the Missouri. Riverside development and public spaces are fostering new physical and cultural interfaces with the river.

In my view, recent floods should lead to more of this kind of rethinking. River towns can start by restricting floodplain development so that people and property will not be in harm's way. This will create space for rivers to spill over in flood season, reducing risks downstream. Proposals to raise and improve levees should be required to take climate change and related flooding risks into account.

Davenport, Iowa, has embraced this approach. With a

population of over 102,000, it is the largest US river city without levees or a permanent floodwall. Instead Davenport has emphasized adapting to flooding by increasing public green spaces in the flood zone and elevating buildings that flank the Mississippi River.

Kansas City and other towns could advance this discussion by moving beyond strictly commercial visions of their waterways and considering this question: What does a healthy river town of the future look like?

Notes

1. Mitch Smith and John Schwartz, "'Breaches Everywhere': Flooding Bursts Midwest Levees, and Tough Questions Follow," *New York Times*, March 31, 2019.
2. Donald Worster, *Rivers of Empire: Water, Aridity, and the Growth of the American West* (New York: Oxford University Press, 2010).
3. L. Fox, "25 Years and Millions of Dollars after the Great Flood, Is Kansas City Safer?," *Kansas City Star*, July 29, 2018, https://www.kansascity.com/news/politics-government/article215578845.html.
4. Samuel E. Munoz, Liviu Giosan, Matthew D. Therrell, Jonathan W. F. Remo, Zhixiong Shen, Richard M. Sullivan, Charlotte Wiman, Michelle O'Donnell, and Jeffrey P. Donnelly, "Climatic Control of Mississippi River Flood Hazard Amplified by River Engineering," *Nature* 556, no. 7699 (2018): 95–98, https://doi.org/10.1038/nature26145.
5. William Least Heat-Moon, *River-Horse: The Logbook of a Boat across America* (New York: Penguin, 1999).

Part IV.

The Lifeblood of Human Society

For all of humanity's existence, people have settled near water. To better understand the importance of water to society, part IV gathers contributions by environmentalists, sociologists, policy experts, and legal scholars that examine the culture that surrounds water, the conflict that water can bring, and the centrality of water in human life and well-being. The contributors reveal the rich relationships that many communities, especially Indigenous communities, have to water and the cultural values at risk in a changing climate. LaPier tells of the efforts of environmental groups fighting to protect rivers and landscapes based on their "personhood." In New Zealand, for instance, the national government recognized the ancestral connection of the Māori people to their water by according personhood status to the Whanganui River in 2017, thereby providing it with the same rights and powers of a legal person.

According to Feldman, the COVID-19 pandemic shed light on the public health challenges associated with a lack of access to clean water for people in many parts of the world, further pressing the need to rebuild public trust and address today's water challenges. As governments invest in infrastructure and respond to climate change, issues of local culture and landscape attachment come to the fore. Despite the significant planning and infrastructure investment under way in Louisiana, for example, planners are failing to protect and restore culture disrupted by land loss or by new restoration projects. Without community-wide resettlement assistance, explains Colten, the coastal region's distinctive cultures are at risk of community fragmentation and cultural dissolution.

Gender disparities associated with water are prominent in many developing regions where women spend a significant amount of their time hauling water to their household and caring for family members who get ill from consuming poor-quality water. This time spent means lost opportunities for women's employment, education, leisure, and even their sleep, according to Caruso. There is an urgency to engage women and girls in designing development programs to improve access to water if we expect these programs to succeed.

We are also better understanding the public health benefits associated with living close to water, or accessing blue space, as it is sometimes called. Contributors Georgiou and Chastin are discovering in their research that having more blue space—in the form of still and running freshwater—in neighborhoods can significantly increase peoples' physical activity levels, lower their stress and anxiety, and enhance their overall psychological well-being. New projects under way in the United Kingdom are regenerating canal networks to create new blue spaces. In addition to benefiting public health, these projects can help control water levels and mitigate flooding, therefore making communities more resilient to climate change.

The chapters of part IV, taken together, bring more of a human focus to water, revealing how it is the lifeblood of human society.

Chapter 23

For Native Americans,
a River Is More Than a "Person";
It Is Also a Sacred Place

ROSALYN R. LaPIER

IN 2017, the environmental group Deep Green Resistance filed
a first-of-its-kind legal suit against the State of Colorado,
asking for personhood rights for the Colorado River. They
argued that the river had a right to "exist."[1] The case was ultimately dismissed for technical reasons. In the past, environmental groups in India, Bolivia, Ecuador, Colombia, and New
Zealand have successfully sought protection for rivers and
landscapes based on this argument.

    As a Native American scholar of environment and religion,

I seek to understand the relationship between people and the natural world. Native Americans view nature through their belief systems. A river or water does not only sustain life—it is also sacred.

Why Is Water Sacred to Native Americans?

In recent years, the Lakota phrase *Mní wičhóni*, or "Water is life," has become a new national protest anthem. It was chanted by thousands of marchers at the Treaty People Gathering to oppose Line 3 in Minnesota in the summer of 2021, and it originated at protests as the anthem of the struggle to stop the building of the Dakota Access Pipeline under the Missouri River in North Dakota.

There was a reason: for long years, the Lakota, the Blackfeet, and other Native American tribes understood how to live with nature, an understanding that came from living within the restrictions of the limited water supply of the "Great American Desert" of North America.[2]

Water as Sacred Place

Native Americans learned both through observation and experiment, arguably a process similar to what we call science today. They also learned from their religious ideas, passed on from generation to generation in the form of stories.

I learned from my grandparents, both members of the Blackfeet tribe in Montana, about the sacredness of water. They shared that the Blackfeet believed in three separate realms of existence—the Earth, sky, and water. The Blackfeet believed that humans, or "Niitsitapi," and Earth beings, or "Ksahkomitapi," lived in one realm; sky beings, or "Spomitapi," lived in another realm; and underwater beings, or "Soyiitapi," lived in yet another. The Blackfeet viewed all three worlds as sacred because within them lived the divine.

The water world, in particular, was held in special regard. The Blackfeet believed that in addition to the divine beings, about which they learned from their stories,[3] there were divine animals. The divine beaver, who could talk to humans, taught the Blackfeet their most important religious ceremony. The Blackfeet needed this ceremony to reaffirm their relationships with the three separate realms of reality.

The Soyiitapi, divine water beings, also instructed the Blackfeet to protect their home, the water world. The Blackfeet could not kill or eat anything living in water; they also could not disturb or pollute water. The Blackfeet viewed water as a distinct place—a sacred place. It was the home of divine beings and divine animals who taught the Blackfeet religious rituals and moral restrictions on human behavior. It can, in fact, be compared to Mount Sinai of the Old Testament, which was viewed as "holy ground" where God gave Moses the Ten Commandments.

Water as Life

Native American tribes on the Great Plains knew something else about the relationship among themselves, the beaver, and water. They learned through observation that beavers helped create an ecological oasis within a dry and arid landscape.

As Canadian anthropologist R. Grace Morgan hypothesized in her book *Beaver, Bison, Horse*, the Blackfeet sanctified the beaver because they understood the natural science and ecology of beaver behavior.[4] Morgan believed that the Blackfeet did not harm the beaver because beavers build dams on creeks and rivers. Such dams could produce enough of a diversion to create a pond of fresh, clean water that allowed an oasis of plant life to grow and wildlife to flourish. Beaver ponds provided the Blackfeet with water for daily life. The ponds also attracted animals, which meant the Blackfeet did not have to travel long distances to hunt. The Blackfeet did not need to travel for plants used for medicine or food, either.

Beavers were part of what ecologists call a trophic cascade, or a reciprocal relationship. Beaver ponds were a win-win for all concerned in the Great American Desert, which modern ecologists and conservationists are beginning to study only now. For the Blackfeet, Lakota, and other tribes of the Great Plains, water was "life." They understood what it meant to live in a semiarid place, which they expressed through their religion and within their ecological knowledge.

Rights of Rivers

Indigenous people from around the world share these beliefs about the sacredness of water.

The government of New Zealand has begun to recognize the ancestral connection of the Māori people to their water. In the spring of 2017, the government passed the Te Awa Tupua Whanganui River Claims Settlement Bill, which provides "personhood" status to the Whanganui River, one of the largest rivers on the North Island of New Zealand.[5] This river has come to be recognized as having "all the rights, powers, duties, and liabilities of a legal person"—something the Māori believed all along.

The United States does not have such "personhood" laws. Indigenous people recognize that a river is more than a person—it is also a sacred place.

Notes

1. The Colorado River Ecosystem, Deep Green Resistance, The Southwest Coalition, Deanna Meyer, Jennifer Murnan, Fred Gibson, Susan Hyatt, Will Falk, Plaintiff, v. State of Colorado, Defendant. Case 1:17-cv-02316. US District Court for the District of Colorado. Filed September 25, 2017. https://www.courthousenews.com/wp-content/uploads/2017/09/Colorado-River.pdf.

2. Map of the country embracing the route of the expedition of 1823 commanded by Major S. H. Long. Digitized image. Michigan State University Libraries, https://lib.msu.edu/branches/maps/MSU-Scanned/North_America_Canada/800-c-reg4-D-1823-400/. See also a map of the Great American Desert, drawn by Stephen H. Long in 1820, available as an image from Wikipedia, https://en.wikipedia.org/wiki/Great_American_Desert#/media/File:Great_American_Desert,_mapped_by_Stephen_H._Long.jpg.

3. Rosalyn R. LaPier, *Invisible Reality: Storytellers, Storytakers, and the Supernatural World of the Blackfeet* (Lincoln: University of Nebraska Press, 2019).

4. R. Grace Morgan, *Beaver, Bison, Horse: The Traditional Knowledge and Ecology of the Northern Great Plains* (Regina, SK: University of Regina Press, 2020).

5. Te Awa Tupua (Whanganui River Claims Settlement) Bill. Government Bill 129–2. New Zealand Legislation (2017), https://www.legislation.govt.nz/bill/government/2016/0129/latest/DLM6830851.html?src=qs.

Chapter 24

Louisiana's Coastal Cultures Are Threatened by the Very Plans Meant to Save Their Wetlands and Barrier Islands

CRAIG E. COLTEN

WAVES OF DISASTER have earned Louisiana a reputation as the place to watch for how climate change will impact coastal areas. Hurricane Ida in 2021 was merely a punctuation mark in a series of devastating tropical cyclones, tragic inland floods, epic oil spills, and deadly epidemics. Despite these all-too-frequent catastrophes, many residents of Louisiana's vulnerable coastal areas remain firmly committed to rebuilding after each disaster. The powerful pulls of family, faith, traditional foods, local music, culture, and landscapes create a strong attachment.

Native Americans, African Americans, Acadians, Isleños, and Vietnamese populate the coastal region, living in narrow settlements along the bayous and natural levees that stand a few feet above the backwater swamps and marshes. Many come from a history of traumatic displacement from their traditional homelands. They adapted to the local environment; became sugarcane farmers, shrimpers, fishers, or oyster farmers; and sunk deep roots.

In coastal Louisiana, people often live their entire lives near where they were born. Yet they have also moved, incrementally "up the bayou"—away from the Gulf of Mexico—over the decades in order to survive in a perilous place.[1] Each major storm prompts a few more departures that contribute to a slow trickle of recovery-weary residents.

As the state tries to cope with repeat catastrophes, it is figuring out how to manage an ongoing crisis: the slow-motion loss of these southern wetlands and barrier islands. They provide valuable natural storm protection. But the state's solutions may end up harming the communities that live there and endangering the unique cultures that define the Louisiana coast.

As a historical geographer living in Louisiana, I study these areas and published a book on Louisiana's land-loss crisis.[2] My research documents how these rural residents are being asked to adapt to save cities and industries and how that's affecting their cultures.

The Downside to Wetlands Restoration

The state's coastal margins have been disappearing at the rate of about 23 square miles per year. That's due in part to flood protection levees that redirected water-borne sediment away from the Mississippi River Delta. This sediment once seasonally rejuvenated the river's floodplain, backswamps, and marshes during spring flooding. Now, it's channeled between high levees, so that material is carried offshore. Without regular replenishment, the delta sinks.[3] Navigation canals dug for oil and gas development have contributed to salt-water intrusion and erosion, furthering land loss. Pumping oil and gas also accelerates the land's subsidence.

The gradual rise of the water level in the Gulf of Mexico as the climate warms, combined with these other processes, exposes Louisiana to the highest rates of relative sea level rise in the United States. That makes the low-lying coastal parishes more susceptible to inundation and storm surge flooding like Ida's.

Fixing One Problem, Creating Another

To offset this slow-moving disaster, the state has launched an ambitious program to fortify the coast and restore wetlands and barrier islands. The plan includes structures to divert Mississippi River water and sediment into the marshes again. But those freshwater diversions bring another problem: they can change the water chemistry and add sediment, affecting the oysters, shrimp, crabs, and fish that residents depend on.

The state's Coastal Protection and Restoration Authority, which is directing this gargantuan effort, is attentive to protecting the major industries and largest cities, restoring critical coastal habitats and ecological functions, and assisting coastal residents. Toward these ends it has spent millions of dollars studying the geology, hydrology, and ecology of the region. And it intends to spend billions on its projects, which would create multiple layers of defense such as restored wetlands and barrier islands, along with levees.

Its regularly updated plans note that local culture matters as well. Yet it hasn't measured the social and cultural processes at work or modeled their future. Planners have offered no designs for protecting and restoring cultures that will be disrupted by either land loss or the projects on the drawing boards.

Cultures at Risk

Distinctive ethnic and cultural groups have persisted here despite living amid the waves of calamity that wash over their homes. Our studies explain how locally based practices have enabled them to rebound, rebuild, and recover after hurricanes, river floods, epidemics, and oil spills.[4] Social scientists refer to these as inherent or informal resilience. Long before the arrival of Civil Defense, FEMA (Federal Emergency Management Agency), or other government-organized response efforts, residents deployed these practices, enabling people reeling from a hurricane to begin rescuing, sheltering, and feeding neighbors and repairing housing and workplaces.

The state's restoration plan neglects these fundamental cultural skills.

The plan allows for "voluntary acquisition" of the homes of those who live beyond the structural protections and wish to depart. Yet there has been no meaningful discussion, study, or planning for assisted resettlement of at-risk communities by the agency in charge of coastal restoration. Another agency has worked for several years to assist the largely Native American community of Isle de Jean Charles to begin an inland move.[5] There is no comparable effort for other communities within the master plan.

Buyouts may enable some families to escape a precarious

situation. But without community-wide resettlement assistance, it will inevitably contribute to community fragmentation and cultural dissolution as residents drift apart. As cultural communities erode after incremental departures following massive storms and other disasters, the state is abetting, unintentionally, the disintegration of the coastal region's distinctive and highly valued cultures.

A Warning to Other Coastal Areas

Louisiana's landscape offers a preview of what might be expected in other locations facing sea level rise and seeking protection behind fixed dikes or levees. These barriers tend to disrupt local environments that resource-based economies such as fishing depend on. They also contribute to a "levee effect": the creation of a false sense of security that exposes coastal residents to severe impacts when a storm exceeds the levee's design limits.

With each successive storm, recovery funds will go into repairing damaged rigid coastal protection systems, like the

Rising seas had already forced many people out of Isle de Jean Charles, a largely Native American community, by the time Hurricane Ida hit in 2021.
AP Photo / Gerald Herbert

US$14 billion to repair the New Orleans levees after Hurricane Katrina and the uncalculated damage to restoration projects caused by Ida. That means less money will be available to address the needs of threatened cultural communities.

Designing protection systems that incorporate informal resilience, such as community-directed resettlement planning, or that integrate with existing social networks can protect both coastal cultures and inland populations. And when conditions become untenable, as some Louisiana settlements are discovering, the state's investments may have to go beyond individual buyouts to help communities plan a safer future together.

Notes

1. Craig E. Colten, Jessica R. Z. Simms, Audrey A. Grismore, and Scott A. Hemmerling, "Social Justice and Mobility in Coastal Louisiana, USA," *Regional Environmental Change* 18, no. 2 (2017): 371–83, https://doi.org/10.1007/s10113-017-1115-7.
2. Craig E. Colten, *State of Disaster: A Historical Geography of Louisiana's Land Loss Crisis* (Baton Rouge: Louisiana State University Press, 2021).
3. T. E. Törnqvist, K. L. Jankowski, Y.-X. Li, and J. L. González, "Tipping Points of Mississippi Delta Marshes due to Accelerated Sea-Level Rise," *Science Advances* 6, no. 21 (2020): https://doi.org/10.1126/sciadv.aaz5512.
4. Craig E. Colten, Jenny Hay, and Alexandra Giancarlo, "Community Resilience and Oil Spills in Coastal Louisiana," *Ecology and Society* 17, no. 3 (2012): https://doi.org/10.5751/es-05047-170305.
5. Jessica R. Z. Simms, Helen L. Waller, Chris Brunet, and Pamela Jenkins, "The Long Goodbye on a Disappearing, Ancestral Island: A Just Retreat from Isle de Jean Charles," *Journal of Environmental Studies and Sciences* 11, no. 3 (2021): 316–28, https://doi.org/10.1007/s13412-021-00682-5.

Chapter 25

Women Still Carry
Most of the World's Water

BETHANY CARUSO

IMAGINE GOING THROUGH YOUR DAY without access to clean, safe water in your home for drinking, cooking, washing, or bathing whenever you needed it. According to a 2021 report from United Nations Children's Fund (UNICEF) and the World Health Organization (WHO), two billion people around the world face that challenge every day.[1] And the task of providing water for households falls disproportionately to women and girls, especially in rural areas.[2]

Water, a human right, is critical for human survival and development. A sufficient supply of biologically and chemically safe water is necessary for drinking and personal hygiene to prevent diarrheal diseases, trachoma, intestinal worm infections, stunted growth in children, and numerous other deleterious outcomes from chemical contaminants like arsenic and lead.

I have carried out research in India, Bolivia, and Kenya on the water and sanitation challenges that women and girls confront and how these experiences influence their lives. In my fieldwork I have seen adolescent girls, pregnant women, and mothers with small children carrying water. Through interviews I learned of the hardships they face when carrying out this obligatory task. An insufficient supply of safe and accessible water poses extra risks and challenges for women and girls. Without recognizing the uneven burden of water work that women bear, well-intentioned programs designed to bring water to places in need will continue to fail in meeting their goals.

Lost Hours

Collecting water takes time. In fetching water for drinking, bathing, cooking, and other household needs, millions of women and girls spend hours every day traveling to water sources, waiting in line, and carrying heavy loads—often several times a day.

The 2021 UNICEF and WHO report states that 282 million people worldwide have access to water sources that are considered safe but need to spend at least 30 minutes walking or queuing to collect their water. Another 122 million get their water from surface sources that are considered to be the most unsafe, such as rivers, streams, and ponds. Water from these sources is even more likely to require over 30 minutes to collect.

In a study of 25 countries in sub-Saharan Africa, UNICEF estimated that women there spent 16 million hours collecting water each day. Women in a study in Kenya reported spending an average of 4.5 hours fetching water per week, causing 77 percent to worry about their safety while fetching and preventing 24 percent from caring for their children.

When children or other family members get sick from consuming poor-quality water, which can happen even if the water is initially clean when collected, women spend their time providing care. These responsibilities represent lost opportunities for women's employment, education, leisure, or sleep. Moreover, significant disruptions to water services due to COVID-19 have been reported, adding further complication.

Heavy Loads

Water is heavy. The WHO recommends 20–50 liters of water per person per day for drinking, cooking, and washing. That amounts to hauling between 44 and 110 pounds of water daily for use by each household member.

And in many places, water sources are far from homes. In Asia and Africa, women walk an average of 6 kilometers (3.7 miles) per day collecting water.[3] Carrying such loads over long distances can result in strained backs, shoulders, and necks and in other injuries if women have to walk over uneven and steep terrain or on busy roads. The burden is even heavier for women who are pregnant or are also carrying small children. Moreover, pregnant women worry that transporting these heavy loads will lead to early labor or even miscarriage.

Even when a household or village has access to a safe water source close to home, residents may not use it if they believe the water is inferior in some way. As one woman told

my research team in India: "Tube well water quality is not good
. . . water is saline. Cooking is not good due to this water. Not
good for drinking either. People are getting water from that
neighboring village. . . . For cooking we get water from the
river." In this community, the neighboring village was at least a
kilometer away.

Fetching water can also be dangerous for the personal
safety of women and girls. They may encounter conflict at
water points or the risk of physical or sexual assault. Many of
these same dangers also arise when women do not have access to safe, clean, and private toilets or latrines for urinating,
defecating, and managing menstruation.

Global demand for water is increasing. The United
Nations forecasts that if current water use patterns do not
change, world demand will exceed supply by 40 percent by
2030. In such a scenario, it is hard to imagine that women's
and girls' experiences will improve without intentional efforts.

A Focus on Women's Needs

When communities initiate programs to improve access to
water, it is critical to ask women about their needs and experiences. Although women and girls play key roles in obtaining
and managing water globally, they are rarely offered roles in
water improvement programs or on local water committees.
They need to be included as a right and as a practical matter.
Numerous water projects in developing countries have failed
because they did not include women.

And women should play meaningful roles. A study in
northern Kenya found that although women served on local
water management committees, conflict with men at water
points persisted because the women often were not invited

to meetings or were not allowed to speak.[4] Women who raise their voices about water concerns need to be heard. In Flint, Michigan, women were critical to revealing the city's water crisis and continue to push for changes.

We also need broader strategies to reduce gender disparities associated with accessing water. First, we need to collect more data on women's water burden and how it affects their health, well-being, and personal development.[5] Second, social messaging affirming the idea that water work belongs only to women must be abandoned. Third, women must be involved in creating and managing targeted programs to mitigate risks and simultaneously work to improve women's lives and enhance their empowerment.[6] Finally, these programs should be evaluated to determine whether they are truly improving women's lives.

Ban Ki-moon, former United Nations secretary-general, has called empowering the world's women "a global imperative." To attain that goal, we must reduce the weight of water on women's shoulders.

Notes

1. *Progress on Household Drinking Water, Sanitation and Hygiene, 2000–2020*, WHO/UNICEF Joint Monitoring Programme for Water Supply, Sanitation, and Hygiene, 2021, https://washdata.org/sites/default/files/2022-01/jmp-2021-wash-households_3.pdf.

2. Susan B. Sorenson, Christiaan Morssink, and Paola Abril Campos, "Safe Access to Safe Water in Low Income Countries: Water Fetching in Current Times," *Social Science & Medicine* 72, no. 9 (2011): 1522–26, https://doi.org/10.1016/j.socscimed.2011.03.010; Jay P. Graham, Mitsuaki Hirai, and Seung-Sup Kim, "An Analysis of Water Collection Labor among Women and Children in 24 Sub-Saharan African Countries," *PLOS One* 11, no. 6 (2016): https://doi.org/10.1371/journal.pone.0155981.

3. *The Right to Water: Fact Sheet No. 35*. United Nations Human Rights Office of the High Commissioner, https://www.ohchr.org/Documents/Publications/FactSheet35en.pdf.

4. Sarah Yerian, Monique Hennink, Leslie E. Greene, Daniel Kiptugen,

Jared Buri, and Matthew C. Freeman, "The Role of Women in Water Management and Conflict Resolution in Marsabit, Kenya," *Environmental Management* 54, no. 6 (2014): 1320–30, https://doi.org/10.1007/s00267-014-0356-1.

5. Bethany A. Caruso and Sheela S. Sinharoy, "Gender Data Gaps Represent Missed Opportunities in WASH," *Lancet Global Health* 7, no. 12 (2019): https://doi.org/10.1016/s2214-109x(19)30449-8.

6. Bethany A. Caruso, Amelia Conrad, Madeleine Patrick, Ajilé Owens, Kari Kviten, Olivia Zarella, Hannah Rogers, and Sheela S. Sinharoy, "Water, Sanitation, and Women's Empowerment: A Systematic Review and Qualitative Metasynthesis," *PLOS Water* 1, no. 6 (2022): e0000026, https://doi.org/10.1371/journal.pwat.0000026.

Chapter 26

Coronavirus Spotlights the Link between Clean Water and Health

DAVID FELDMAN

OVER THE COURSE of the coronavirus pandemic, experts have told us that a key way to minimize the odds of getting sick is by washing hands thoroughly and frequently and by cleaning surfaces with which we come into contact. But what if you don't have access to clean water?

Over the past 40 years, many nations have made great progress in treating wastewater, providing residents with clean drinking water, and enhancing water supplies to grow needed food and fiber. But as a researcher focusing on water resources management and policy, I know there is still far to go. More than 40 percent of the world's population lives in regions where water is becoming increasingly scarce, and that figure is likely to rise. Every day, nearly 1,000 children die from preventable water- and sanitation-related diseases.[1]

Life without Clean Water

Water use has increased worldwide by about 1 percent annually since the 1980s, driven by population growth, economic development, and changing consumption patterns.[2] At the same time, water supplies are increasingly threatened by climate change, overuse, and pollution.

For example, in 2019 residents of Chennai, India, had to queue up for water delivered by tanker trucks because the city's reservoirs were empty. Persistent drought, worsened by climate change, had virtually exhausted local supplies. The city, which is home to seven million people, still faces severe shortages and may exhaust its available groundwater within a few years.

In rural Mexico, some five million people lack access to clean water.[3] Women and children are tasked with collecting water, taking time that could be spent in school or on political engagement. Meanwhile, men decide how water rights are allocated.

Residents of Flint, Michigan, whose trust in the safety of their drinking water has been gradually restored after a notorious case of lead contamination, were advised in August 2019 to boil water as a precaution against impurities after a pipeline rupture reduced pressure in the city's water lines. The advisory ended after sampling indicated that there was no danger of contamination, but the city is still replacing lead and galvanized steel water-delivery pipes to prevent further lead exposure.

Washing hands is a difficult challenge in many developing countries. Clean water and soap are often in short supply, and many slum dwellers live in homes without running water.[4]

Systems under Stress

According to the United Nations, rising demand for water in the industrial, domestic, and agricultural sectors signals that people are starting to live better, thanks to progress in harnessing freshwater for growing food and fiber and for public consumption. However, experts note three areas where progress is lagging.

First, more than two billion people live in countries experiencing high water stress, and about four billion people experience severe water scarcity during at least one month of the year. These problems are directly attributable to rising water demands and the intensifying effects of climate change. They also worsen mistreatment of women, who bear much of the burden of providing scarce water to families.

Second, while many countries are spending money on improving access to water—often by privatizing supplies, which enriches the multinational engineering companies that build the infrastructure—the supply of clean water remains inadequate. Nearly 800 million people worldwide lack updated sanitation. In many instances primitive latrines release human wastes directly to the environment, contaminating streams and rivers. Worldwide, over 80 percent of wastewater from human activities remains untreated.

Third, in every country, water infrastructure is deteriorating, and people are disposing of drugs, personal care products, and other common household goods into public water systems. These combined trends add persistent, hard-to-treat contaminants to water supplies and threaten public health worldwide.

Water as a Leadership Test

These problems are daunting, but progress is possible if water agencies and government officials engage the public, heed evidence-based advice from experts, and exercise political leadership.

As a first step, governments need to focus on long-term planning and coordinated responses. The problems facing Chennai, rural Mexico, Flint, and countless other places usually generate early warning signs, which public officials often ignore due to a lack of political will or sense of urgency. In Cape Town, South Africa, where residents faced a water shortage in 2017 similar to Chennai's, it had been clear for years that the city's water infrastructure could not handle growing demands. A government-sponsored study published in 1998 had recommended building a wastewater reuse plant as a hedge against future drought, but the plant was never constructed. Flint's water crisis escalated over some 18 months while public officials closed their ears to residents' frequent complaints about the smell and taste of their water. The good news is that many large cities, including Los Angeles and São Paulo, Brazil, have begun to heed warning signs of climate change. In response, public officials are initiating innovative water alternatives that conserve water, reuse wastewater, and harvest rainwater.

Second, it is important to recognize water problems as environmental justice challenges. The United Nations' International Hydrological Program now promotes water equity, recognizing that the burdens of protracted drought, water stress, and contaminated supplies fall disproportionately on women, the young, the frail and destitute, and oppressed Indigenous minorities, who often are forced to migrate elsewhere when conditions become intolerable.[5] Here in the United States, many cities and states passed laws forbidding water agencies from cutting off water supplies to households unable to pay their bills during the pandemic. Unfortunately, many of these measures were short-term, and when they expired, millions faced the renewed challenge of paying water bills.

Finally, I believe that building or restoring public trust is critical for addressing these problems. The experience of cities that have weathered drought, such as Melbourne, Australia, shows that governments need to weigh and address community concerns and to foster trust and confidence in the agencies charged with implementing solutions. In my view, the best way to build that kind of trust is by courageously meeting today's water crises head-on.

Notes

1. "Goal 6: Ensure Access to Water and Sanitation for All," United Nations Sustainable Development Goals, https://www.un.org /sustainabledevelopment/water-and-sanitation/.
2. *UN World Water Development Report 2019*, UN Water, March 18, 2019, https://www.unwater.org/publications/world-water-development-report -2019/.
3. *Case Study: Water and Sanitation Management with a Gender Perspective in Mexico*, Sustainable Development Goals Fund, May 14, 2017, https://www.sdgfund.org/case-study/water-and-sanitation -management-gender-perspective-mexico.
4. Swaminathan Natarajan, "Coronavirus: Why Washing Hands Is Difficult in Some Countries," BBC News, March 18, 2020, https://www.bbc.com /news/world-51929598.
5. "Climate Change and Disaster Displacement," UNHCR, the UN Refugee Agency, https://www.unhcr.org/en-us/climate-change-and-disasters .html.

Chapter 27

Living near Water Can Be Beneficial to Your Mental Health— Here's How to Have More Blue Spaces in Cities

MICHAIL GEORGIOU and SEBASTIEN CHASTIN

WE KNOW THAT PEOPLE are more likely to experience mental health disorders in areas with greater population density. Overcrowding, pollution, urban violence, and less social support may all be contributing factors, and this is becoming more of a challenge as more people around the world move into towns and cities.[1]

Natural settings have long been seen as a potential solution. Many studies have shown that when people are closer to nature, they are less stressed, and their mood and general mental health improve.[2] There has been much research into using therapeutic landscapes in cities to bring the benefits of being in nature to more people.[3]

While plenty of studies have focused on green spaces, researchers are also beginning to look at blue spaces to learn about the health benefits of living near water. So far, studies show that people living near water have a lower risk of premature death, have a lower risk of obesity, and generally report better mental health and well-being. Blue spaces also reduce the gap between less and more affluent areas in the risk of dying prematurely. Being close to water does enhance people's well-being, but no research has yet shown that it reduces the incidence of mental health disorders. Moreover, most studies have focused on coastal towns rather than cities.

Given that even landlocked cities are built around water features like canals, rivers, and lakes, our research aimed to uncover their health benefits and to see how they could be repurposed to improve the mental health of people living in cities.

Mental Health Connection

We conducted a systematic review and meta-analysis of all the evidence about how blue space positively impacts health.[4] This showed that living closer to and having more blue space within a neighborhood could significantly increase residents' physical activity levels. Blue spaces were also shown to lower stress and anxiety, while boosting people's mood and psychological well-being. Our findings align with what other studies have found.[5]

Researchers studying the effects of blue space delivered through virtual reality have also found that people see it as restorative, fascinating, and preferable to a built-up environment.[6] This shows how technology could be used as a way of studying how being near water affects people.

Our next area of research will be to understand how blue spaces benefit people in these ways. We also think that having more stretches of water in cities could improve the population's health in other ways, such as by reducing heat and lowering air pollution. But more research will be needed to understand whether this is true.

Creating Blue Spaces

If the early evidence points to lots of health benefits from living near water, the problem many cities encounter is finding ways to bring them to residents.

During the Victorian era, for example, canals in the United Kingdom were tremendously important to the economy. Canals allowed trade to happen and helped workers move around. There's still a huge network of these waterways in many cities in the United Kingdom, but few of them are in use. There are more canals in Birmingham, for instance, than in Venice. But access to them is often blocked by tall buildings or fences, and their potential is far from being exploited. Derelict canals can sometimes even cause environmental problems, such as by being a site of plastic litter pollution, which can reduce biodiversity and harm wildlife.

Numerous projects in the last few years have sought to regenerate canal networks in the United Kingdom, though mainly with a view to improving the local economy by creating valuable real estate. But even aside from the potential mental health effects, regenerating these networks can bring other benefits, such as controlling water levels, preventing floods, and making cities more resilient to climate change. With this in mind, some municipalities are starting to use canals to mitigate flood risks and to provide greener transport options. This is creating a win-win-win situation that combines economic, environmental, and health benefits.

Efforts in Scotland to regenerate canals are a good example. Land that was previously flooded by rainwater has seen new homes and businesses built. Walking paths along the canals have also been created, allowing people to visit the canals, even while further development is under way.

Further research is yet needed to better understand the true benefits that blue spaces have for residents; it is clear, though, that finding ways to repurpose derelict canals in cities could have other benefits for the environment and economy too.

Notes

1. Kalpana Srivastava, "Urbanization and Mental Health," *Industrial Psychiatry Journal* 18, no. 2 (2009): 75–76, https://doi.org/10.4103/0972-6748.64028.
2. Magdalena van den Berg, Wanda Wendel-Vos, Mireille van Poppel, Han Kemper, Willem van Mechelen, and Jolanda Maas, "Health Benefits of Green Spaces in the Living Environment: A Systematic Review of Epidemiological Studies," *Urban Forestry & Urban Greening* 14, no. 4 (2015): 806–16, https://doi.org/10.1016/j.ufug.2015.07.008.
3. Wilbert M. Gesler, "Therapeutic Landscapes: Theory and a Case Study of Epidauros, Greece," *Environment and Planning D: Society and Space* 11, no. 2 (1993): 171–89, https://doi.org/10.1068/d110171.
4. Michail Georgiou, Gordon Morison, Niamh Smith, Zoë Tieges, and Sebastien Chastin, "Mechanisms of Impact of Blue Spaces on Human Health: A Systematic Literature Review and Meta-analysis," *International Journal of Environmental Research and Public Health* 18, no. 5 (2021): 2486, https://doi.org/10.3390/ijerph18052486.
5. Sebastian Völker and Thomas Kistemann, "The Impact of Blue Space on Human Health and Well-Being—Salutogenetic Health Effects of Inland Surface Waters: A Review," *International Journal of Hygiene and Environmental Health* 214, no. 6 (2011): 449–60, https://doi.org/10.1016/j.ijheh.2011.05.001.

6. Giovanna Calogiuri, Sigbjørn Litleskare, Kaia A. Fagerheim, Tore L. Rydgren, Elena Brambilla, and Miranda Thurston, "Experiencing Nature through Immersive Virtual Environments: Environmental Perceptions, Physical Engagement, and Affective Responses during a Simulated Nature Walk," *Frontiers in Psychology* 8 (2018): https://doi.org/10.3389/fpsyg.2017.02321.

Part V.

Preserving Our Oceans

In previous generations, the ocean was seen as an infinitely vast resource uncontrollable by human hands. But as we exhaust resources on land, humans are increasingly turning to the oceans for oil and gas development, aquaculture, tourism, and increased shipping, among other activities. This is stimulating what researchers like Blasiak are calling a Blue Acceleration, which is expected to have consequences for life on the blue planet. Today we know that our oceans are immensely powerful yet also fragile and in need of protection from overfishing, microplastic pollution, and other strains.

The research under way is impressive. Interdisciplinary teams of physicists, biologists, and mathematicians are starting to predict the pathways of different kinds of plastic in our oceans. Finding out where plastic goes in the oceans can help us figure out which parts of the ocean are most affected by it and, ultimately, where to focus cleanup efforts, as Sutherland, DiBenedetto, and van den Bremer explain. O'Connell reflects on her decades of studying our oceans' floors. She reports on the latest technologies and scientific findings from ocean floor drilling and describes how collecting core samples from the deep seabed helps scientists reconstruct Earth's history and understand ocean life that evolved in extreme environments.

To combat threats to our oceans, communities are experimenting with new governance strategies, embracing new technologies, and employing environmental stewardship practices. Establishing marine protected areas is one important regulatory practice, according to Reimer. If designed well, and made attentive to equity concerns, these areas can help maintain healthy ecosystems and sustainable fisheries by defining and limiting activities that can harm ocean life. But so little of the global ocean—only about 6 percent—is protected today.

Indigenous forms of marine management are experiencing a revival in some parts of the world today. In their research on alternative governance models to improve equity and justice for Indigenous peoples, Parsons and Taylor describe how coastal communities, from Fiji to Vanuatu to Samoa, are embracing shared decision-making with governments and are reenacting Indigenous forms of marine management. This involves the stewardship of local fisheries and the creation and implementation of marine protection zones. These governance systems are based on Indigenous knowledge and emphasize reciprocity, connectivity, and holistic approaches.

The chapters of part V provide a big-picture look at what humanity is learning about the health of the world's oceans and about human efforts to protect them for the future.

Chapter 28

How Marine Protected Areas Help Safeguard the Ocean

JULIE REIMER

THE OCEAN QUIETLY UNITES global communities in a pro-
found way. And yet, the ocean faces more threats today than
ever before in history.

The facts and predictions are staggering. More than 25
percent of the world's fisheries are overfished, and most are
operating at a maximum sustainable level with no room for
those fisheries to expand. The rate of ocean acidification,
which happens when the ocean absorbs too much carbon, far
exceeds what is natural.[1] By the turn of the century, more than
half of the world's ocean species could face extinction.

People continue to carelessly destroy the world's biggest ecosystem. Ocean conservation is not just essential to safeguarding biodiversity, but by protecting the ocean, we also protect ourselves.[2] As the ocean deteriorates under climate change, our communities are at risk as well. For some, this could mean watching special coastal places, including their homes, wash away as storms intensify. For others, this could mean losing the fish that support their livelihoods.

A healthy, vibrant ocean supports healthy communities and a healthy planet.[3] We see this connection in the United Nations Sustainable Development Goal for the ocean, SDG 14: Life Below Water. SDG 14 has seven main targets that aim to "conserve and sustainably use oceans, seas and marine resources for sustainable development."

Protecting the World's Ocean

A study from my research shows how marine protected areas (MPAs) are an important tool for achieving ocean sustainability.[4] Today, only 8.1 percent of the global ocean is protected, and only 2.4 percent of it has strong protection against harmful activities, despite scientists calling for strong protection of at least 30 percent.[5]

In 2020 several countries joined the Global Ocean Alliance and agreed to protect 30 percent of the ocean by 2030, a target called 30x30. If we can achieve this, we can halt biodiversity loss, ensure fish and food for all, and maintain a healthy ocean that can cope with climate change.

Not All Marine Protected Areas Are Created Equal

MPAs, in some form, have existed for decades. In the broadest sense, an MPA is an area that is managed to protect some element of ocean life by defining areas where activities are—and are not—allowed. Today they come in all sorts of shapes and sizes, but quality is just as important for MPAs as quantity.

Over the years, the pressure to expand global MPAs has led some governments to protect residual areas that are unused or have lower ecological value to avoid conflict and having to make difficult decisions. Picking the low-hanging fruit may not produce the desired outcome. Our protected areas need to be in the right places, and we must use the right regulations to ensure that MPAs can help us with sustainability.

MPAs can assign different levels of protection to ocean life. Setting a protection level may be influenced by conservation objectives, economic interests, and social and cultural needs in the area. On one end of this spectrum, MPAs can be minimally protected. In these MPAs, some extractive activities are allowed to continue, such as using fishing gear that can inflict damage on species and habitats. There may still be some conservation benefit, albeit a small one. On the other end, MPAs can be highly or fully protected. In these MPAs, only light extraction and destructive activities are permitted (highly protected) or none at all (fully protected), giving our best protection to biodiversity.

On paper, some countries have already surpassed the 30 percent goal, but if MPAs do not have strong regulations, they are not likely to deliver the conservation benefits we need. For example, in France, 33.7 percent of the nation's waters is protected, but less than 2 percent of this area is fully protected.[6] The strongest MPAs have at least some portion of the area fully protected from harmful activities.

On the Atlantic coast of Canada, northern bottlenose whales swim freely in the rich waters of an underwater canyon in an MPA known as "the Gully." As one of Canada's oldest MPAs, this area protects hundreds of species, including whales and dolphins, fish, crabs, octopus, and other invertebrates, not to mention the highest diversity of cold-water corals in Eastern Canada.[7] Located offshore, the Gully might have little local attachment, but fishers use the area, connecting this far-off place to Canadian communities. Despite being out of sight, the Gully is not out of mind.

The Gully was criticized in the past for lax regulations on some activities. Then, in 2019, the Canadian government banned oil and gas activities, mining, and bottom trawling in all of Canada's MPAs. Through its three management zones, including a fully protected zone from the surface to the seafloor, the Gully continues to shine as a Canadian conservation success story.

The Way Forward

MPAs hold the potential to help us achieve some targets of SDG 14. They can help to maintain healthy ecosystems and sustainable fisheries. They are less able to help with other SDG 14 targets, though, like reducing pollution or the impacts of acidification. MPA regulations end at their boundaries, but our harmful impacts do not. MPAs are one type of tool in a larger toolbox that also includes fisheries management, shipping regulations, climate-change policies, and more.

In an ever-changing, increasingly busy ocean, our MPAs have to work with one another and with other tools. Together, they form the bigger picture of ocean management across habitats, ecosystems, and geopolitical boundaries. On our path to protecting a third of the ocean, we must be protecting the right areas, with strong regulations and without exacerbating injustice and inequity in the communities affected by conservation decisions. While today's goal might say 30x30, the real ambition is to see 100 percent of the ocean sustainably managed to ensure a prosperous future for all.

Notes

1. T. Friedrich, A. Timmermann, A. Abe-Ouchi, N. R. Bates, M. O. Chikamoto, M. J. Church, and J. E. Dore, "Detecting Regional Anthropogenic Trends in Ocean Acidification against Natural Variability," *Nature Climate Change* 2, no. 3 (2012): 167–71, https://doi.org/10.1038/nclimate1372.

2. Dan Laffoley, John M. Baxter, Diva J. Amon, Joachim Claudet, Jason M. Hall-Spencer, Kirsten Grorud-Colvert, Lisa A. Levin, et al., "Evolving the Narrative for Protecting a Rapidly Changing Ocean, Post-COVID-19," *Aquatic Conservation: Marine and Freshwater Ecosystems* 31, no. 6 (2021): 1512–34, https://doi.org/10.1002/aqc.3512.

3. Ove Hoegh-Guldberg, Eliza Northrop, and Jane Lubchenco, "The Ocean Is Key to Achieving Climate and Societal Goals," *Science* 365, no. 6460 (2019): 1372–74, https://doi.org/10.1126/science.aaz4390.

4. Julie M. Reimer, Rodolphe Devillers, and Joachim Claudet, "Benefits and Gaps in Area-Based Management Tools for the Ocean Sustainable Development Goal," *Nature Sustainability* 4, no. 4 (2020): 349–57, https://doi.org/10.1038/s41893-020-00659-2.

5. Bethan C. O'Leary, Marit Winther-Janson, John M. Bainbridge, Jemma Aitken, Julie P. Hawkins, and Callum M. Roberts, "Effective Coverage Targets for Ocean Protection," *Conservation Letters* 9, no. 6 (2016): 398–404, https://doi.org/10.1111/conl.12247.

6. Joachim Claudet, Charles Loiseau, and Antoine Pebayle, "Critical Gaps in the Protection of the Second Largest Exclusive Economic Zone in the World," *Marine Policy* 124 (2021): 104379, https://doi.org/10.1016/j.marpol.2020.104379.

7. Maxine C. Westhead, Derek G. Fenton, Tanya A. Koropatnick, Paul A. Macnab, and Hilary B. Moors, "Filling the Gaps One at a Time: The Gully Marine Protected Area in Eastern Canada. A Response to Agardy, Notarbartolo di Sciara and Christie," *Marine Policy* 36, no. 3 (2012): 713–15, https://doi.org/10.1016/j.marpol.2011.10.022.

Chapter 29

Where Does Plastic Pollution Go When It Enters the Ocean?

BRUCE SUTHERLAND, MICHELLE DiBENEDETTO,
and TON VAN DEN BREMER

OF THE HUNDREDS OF MILLIONS OF TONS in plastic waste we produce each year, it's estimated that around 10 million metric tons enter the ocean.[1] Roughly half of the plastics produced are less dense than water, so they float. But scientists estimate that there are only about 0.3 million metric tons of plastic floating at the ocean surface,[2] so where is the rest of it going?

Consider the journey of a plastic fiber that's shed from your fleece. A heavy rain washes it down a storm drain and into a river. Does the tiny fiber settle there? Or does the river carry it out to the coast, where it lingers on the seabed? Or does it continue to float farther out, finally ending up in the vast open ocean?

The dizzying variety of forms that plastic waste can take means that a fiber's fate is just one mystery among countless others. Finding out where all the missing plastic ends up can help us figure out which parts of the ocean are most affected by this type of pollution—and where to focus cleanup efforts. But to do that, we need to be able to predict the pathways of different kinds of plastic, which requires large teams of physicists, biologists, and mathematicians working together.

That's what our research team and collaborators are doing. Here's what we've learned so far.

Plastic Pathways

We know that large pieces of plastic, like bottles, can float on the sea surface for years, if not centuries, taking a long time to break down. Currents, winds, and waves can, after a journey of several years, bring them to the center of ocean basins, where they accumulate in 1,000-kilometer-wide circulating systems known as gyres. The vast "garbage patches" that result resemble more of a soup of plastic than an island of trash.[3]

But the fate of plastic fibers—perhaps the smallest plastic fragments to reach the ocean—is more complex. Large fibers can break up over days and weeks into even smaller pieces, due to turbulence from breaking waves and ultraviolet radiation from the sun. These are called microplastics, and they range in size from 5 millimeters to specks smaller than bacteria. Microplastics can be eaten by fish; it's estimated that one in three fish eaten by humans contains microplastics.

Tinier particles can also be consumed by zooplankton—micro-scopic animals that float at the surface—which are then eaten by even larger animals, including whales.

Microorganisms can grow on the surface of microplastics too, in a process known as "biofouling," which causes them to sink. Muddy rivers, like the Mississippi or the Amazon, contain clays that settle rapidly when they come into contact with salty ocean water. Microplastics can be carried down by the settling clay, but how much this actually happens is unknown.

Quantifying all these outcomes for each bit of plastic is an enormous challenge. What fraction ends up in fish, is carried down by clay, or is covered in microbial slime on the seabed? Of the fraction of plastics that makes it all the way out to the open ocean, it's unclear how long it takes for biofouling or other forces to pull the particles far enough below the surface to start their long descent to the seafloor. With all these complicating factors, it may seem hopeless to predict where plastics ultimately end up. But we're slowly making progress.

Microplastics are often formed from larger
plastic objects that break up in the ocean.
dottedhippo / iStock via Getty Images

Catching a Wave

If you have ever been on a boat in choppy waters, you might
think you're just bobbing up and down in the same spot. But
you're actually moving slowly in the direction of the waves.
This is a phenomenon known as the Stokes drift,[4] and it affects
floating plastics too.

For particles smaller than 0.1 millimeters, moving through
seawater is like us wading through honey. But the viscosity
of seawater has less of an influence on plastics larger than
1 millimeter. Each wave gives these bigger particles a push in
its direction. According to preliminary research that's currently under review, this might mean larger plastics are carried out
to sea much faster than tiny microplastics, making them less
likely to settle in parts of the ocean where more marine life is
found: around coasts.

This research involved studying spherical plastic particles, but microplastic waste comes in all kinds of shapes and
sizes, including disks, rods, and flexible fibers. How do waves
influence where they end up?

A recent study found that nonspherical particles align
themselves with the direction of waves, which can slow the
rate at which they sink.[5] Lab experiments have further shown
how the shape of each plastic particle affects how far it's
transported.[6] Less spherical particles are more likely to travel
farther from coasts.

Solving the mystery of the missing plastics is a science in

its infancy. The ability of waves to transport large microplastics faster than previously thought helps us understand why they are now found throughout the world's oceans, including in the Arctic and around Antarctica. But predicting how far that fleece fiber might travel by water is still more challenging than finding a needle in a haystack.

Notes

1. Jenna R. Jambeck, Roland Geyer, Chris Wilcox, Theodore R. Siegler, Miriam Perryman, Anthony Andrady, Ramani Narayan, and Kara Lavender Law, "Plastic Waste Inputs from Land into the Ocean," *Science* 347, no. 6223 (2015): 768–71, https://doi.org/10.1126/science.1260352.

2. Erik van Sebille, Stefano Aliani, Kara Lavender Law, Nikolai Maximenko, José M. Alsina, Andrei Bagaev, Melanie Bergmann, et al., "The Physical Oceanography of the Transport of Floating Marine Debris," *Environmental Research Letters* 15, no. 2 (2020): 023003, https://doi.org/10.1088/1748-9326/ab6d7d.

3. Erik van Sebille, Matthew H. England, and Gary Froyland, "Origin, Dynamics and Evolution of Ocean Garbage Patches from Observed Surface Drifters," *Environmental Research Letters* 7, no. 4 (2012): 044040, https://doi.org/10.1088/1748-9326/7/4/044040.

4. Ton S. van den Bremer and Øyvind Breivik, "Stokes Drift," *Philosophical Transactions of the Royal Society A: Mathematical, Physical and Engineering Sciences* 376, no. 2111 (2017): 20170104, https://doi.org/10.1098/rsta.2017.0104.

5. Michelle H. DiBenedetto, Jeffrey R. Koseff, and Nicholas T. Ouellette, "Orientation Dynamics of Nonspherical Particles under Surface Gravity Waves," *Physical Review Fluids* 4, no. 3 (2019): https://doi.org/10.1103/physrevfluids.4.034301.

6. Laura K. Clark, Michelle H. DiBenedetto, Nicholas T. Ouellette, and Jeffrey R. Koseff, "Settling of Inertial Nonspherical Particles in Wavy Flow," *Physical Review Fluids* 5, no. 12 (2020): https://doi.org/10.1103/physrevfluids.5.124301.

Chapter 30

Scientists Have Been Drilling into the Ocean Floor for 50 Years— Here's What They've Found So Far

SUZANNE O'CONNELL

IT'S STUNNING BUT TRUE that we know more about the surface of the moon than about the Earth's ocean floor. Much of what we do know has come from scientific ocean drilling: the systematic collection of core samples from the deep seabed. This revolutionary process began over 50 years ago when the drilling vessel Glomar Challenger sailed into the Gulf of Mexico on August 11, 1968, on the first expedition of the federally funded Deep Sea Drilling Project.

I went on my first scientific ocean-drilling expedition in

1980 and since then have participated in six more expeditions to locations including the far North Atlantic and Antarctica's Weddell Sea. In my lab, my students and I work with core samples from these expeditions. Each of these cores, which are cylinders 31 feet long and 3 inches wide, is like a book whose information is waiting to be translated into words. Holding a newly opened core, filled with rocks and sediment from the Earth's ocean floor, is like opening a rare treasure chest that records the passage of time in Earth's history.

Over a half century, scientific ocean drilling has proved the theory of plate tectonics, created the field of paleocean-ography, and redefined how we view life on Earth by revealing an enormous variety and volume of life in the deep marine biosphere. And much more remains to be learned.

Technological Innovations

Two key innovations made it possible for research ships to take core samples from precise locations in the deep oceans. The first, known as dynamic positioning, enables a 471-foot ship to stay fixed in place while drilling and recovering cores, one on top of the next, often in over 12,000 feet of water.

Anchoring isn't feasible at these depths. Instead, techni-cians drop a torpedo-shaped instrument called a transponder over the side. Another device called a transducer, mounted on the ship's hull, sends an acoustic signal to the transponder, which replies. Computers on board calculate the distance and angle of this communication. Thrusters on the ship's hull maneuver the vessel to stay in exactly the same location, countering the forces of currents, wind, and waves.

Another challenge arises when drill bits have to be replaced mid-operation. The ocean's crust is composed of igneous rock that wears bits down long before the desired depth is reached.

When this happens, the drill crew brings the entire drill pipe to the surface, mounts a new drill bit, and returns the pipe to the same hole. This requires guiding the pipe into a funnel shaped reentry cone, less than 15 feet wide, placed at the bottom of the ocean at the mouth of the drilling hole. The process, which was first accomplished in 1970, is like lowering a long strand of spaghetti into a quarter-inch-wide funnel at the deep end of an Olympic-sized swimming pool.

Confirming Plate Tectonics

When scientific ocean drilling began in 1968, the theory of plate tectonics was a subject of active debate. One key idea was that new ocean crust was created at ridges on the seafloor where oceanic plates moved away from each other and magma from earth's interior welled up between them. According to this theory, crust should be new material at the crest of ocean ridges, and its age should increase with distance from the crest.

The only way to prove this was by analyzing sediment and rock cores. In the winter of 1968–1969, the Glomar Challenger drilled at seven sites in the South Atlantic Ocean to the east and west of the Mid-Atlantic Ridge. Both the igneous rocks of the ocean floor and overlying sediments aged in perfect agreement with the predictions, confirming that ocean crust was forming at the ridges and that plate tectonics was correct.

Reconstructing Earth's History

The ocean record of Earth's history is more continuous than geologic formations on land, where erosion and redeposition by wind, water, and ice can disrupt the record. In most ocean locations, sediment is laid down particle by particle, microfossil by microfossil, and remains in place, eventually succumbing to pressure and turning into rock. Microfossils (of plankton) preserved in sediment are beautiful and informative, even though some are smaller than the width of a human hair. As they do with larger plant and animal fossils, scientists can use these delicate structures of calcium and silicon to reconstruct past environments.

Thanks to scientific ocean drilling, we know that after an asteroid strike killed all non-avian dinosaurs 66 million years ago, new life colonized the crater rim within years, and within 30,000 years a full ecosystem was thriving.[1] A few deep-ocean organisms lived right through the meteorite's impact.[2]

Ocean drilling has also shown that 10 million years later, a massive discharge of carbon— probably from extensive volcanic activity,[3] along with methane released from melting methane hydrates—caused an abrupt, intense warming event, or hyperthermal, called the Paleocene-Eocene Thermal Maximum.[4] During this episode, even the Arctic reached more than 73 degrees Fahrenheit.[5] The resulting acidification of the ocean from the release of carbon into the atmosphere and ocean caused massive dissolution and change in the deep-ocean ecosystem.

This episode is an impressive example of the impact of rapid climate warming. The total amount of carbon released during the Paleocene-Eocene Thermal Maximum is estimated to be about equal to the amount that humans will release if we burn all of Earth's fossil fuel reserves. An important difference, though, is that the carbon released by the volcanoes and hydrates was at a much slower rate than we are currently releasing fossil fuel. Thus we can expect even more dramatic climate and ecosystem changes unless we stop emitting carbon.

Finding Life in Ocean Sediments

Scientific ocean drilling has also shown that there are roughly as many cells in marine sediment as in the ocean or in soil.[6] Expeditions have found life in sediments at depths of more than 8,000 feet, in seabed deposits that are 86 million years old, and at temperatures above 140 degrees Fahrenheit.

Today, scientists from 23 nations are proposing and conducting research through the International Ocean Discovery Program, which uses scientific ocean drilling to recover data from seafloor sediments and rocks and to monitor environments under the ocean floor. Coring is producing new information about plate tectonics, such as the complexities of ocean crust formation and the diversity of life in the deep oceans. This research is expensive and is technologically and intellectually intense. But only by exploring the deep sea can we recover the treasures it holds and better understand its beauty and complexity.

Notes

1. Christopher M. Lowery, Timothy J. Bralower, Jeremy D. Owens, Francisco J. Rodríguez-Tovar, Heather Jones, Jan Smit, Michael T. Whalen, et al., "Rapid Recovery of Life at Ground Zero of the End-Cretaceous Mass Extinction," Nature 558, no. 7709 (2018): 288–91, https://doi.org/10.1038/s41586-018-0163-6.
2. V. L. Sharpton and P. D. Ward, "Late Cretaceous–Early Eocene Mass Extinctions in the Deep Sea," in Global Catastrophes in Earth History: An Interdisciplinary Conference on Impacts, Volcanism, and Mass Mortality (Boulder, CO: Geological Society of America, 1990), 481–96.

3. Marcus Gutjahr, Andy Ridgwell, Philip F. Sexton, Eleni Anagnostou, Paul N. Pearson, Heiko Pälike, Richard D. Norris, Ellen Thomas, and Gavin L. Foster, "Very Large Release of Mostly Volcanic Carbon during the Palaeocene–Eocene Thermal Maximum," *Nature* 548, no. 7669 (2017): 573–77, https://doi.org/10.1038/nature23646.

4. James P. Kennett and L. D. Stott, "Abrupt Deep-Sea Warming, Palaeoceanographic Changes and Benthic Extinctions at the End of the Palaeocene," *Nature* 353, no. 6341 (1991): 225–29, https://doi.org/10.1038/353225a0.

5. Appy Sluijs, Stefan Schouten, Mark Pagani, Martijn Woltering, Henk Brinkhuis, Jaap S. Sinninghe Damsté, Gerald R. Dickens, et al., "Subtropical Arctic Ocean Temperatures during the Palaeocene/Eocene Thermal Maximum," *Nature* 441, no. 7093 (2006): 610–13, https://doi.org/10.1038/nature04668.

6. Jens Kallmeyer, Robert Pockalny, Rishi Ram Adhikari, David C. Smith, and Steven D'Hondt, "Global Distribution of Microbial Abundance and Biomass in Subseafloor Sediment," *PNAS* 109 (40) 16213–16, https://doi.org/10.1073/pnas.1203849109.

Chapter 31

Blue Acceleration:
Our Dash for Ocean Resources Mirrors
What We've Already Done on Land

ROBERT BLASIAK

HUMANS ARE LEAVING A HEAVY FOOTPRINT on Earth, but when did we become the main driver of change in the planet's ecosystems?

Many scientists point to the 1950s, when all kinds of socioeconomic trends began accelerating. Since then, the world population has tripled. Fertilizer and water use expanded as more food was grown than ever before. The construction of motorways sped up to accommodate rising car ownership, while international flights took off to satisfy a growing taste for tourism. The scale of human demands on Earth grew beyond historical proportions. This postwar period became known as the "Great Acceleration," and many believe it gave birth to the Anthropocene: the geological epoch during which human activity surpassed natural forces as the biggest influence on the functioning of Earth's living systems.

Researchers studying the ocean are currently feeling a sense of déjà vu. Over the past three decades, patterns seen on land 70 years ago have been occurring in the ocean. We're living through a "Blue Acceleration," and it will have significant consequences for life on the blue planet.[1]

Why Is the Blue Acceleration Happening Now?

As land-based resources have declined, hopes and expectations have increasingly turned to the ocean as a new engine of human development. Take deep-sea mining. The international seabed with its mineral riches has excited commercial interest in recent years due to soaring commodity prices. According to the International Monetary Fund, the price of gold is up 454 percent from 2000, silver up 317 percent, and lead 493 percent. Around 1.4 million square kilometers of the seabed have been leased by the International Seabed Authority since 2001 for exploratory mining activities.

In some industries, technological advances have driven these trends. Virtually all offshore wind farms were installed in the last 20 years. The marine biotechnology sector scarcely existed at the end of the 20th century, and over 99 percent of the genetic sequences from marine organisms found in patents were registered after 2000.[2]

During the 1990s, as the Blue Acceleration got under way, the world population reached 6 billion. Today there are around 7.8 billion people. Population growth in water-scarce areas like the Middle East, Australia, and South Africa has caused a threefold growth in the volume of desalinated seawater generated since 2000. Increasing population has also meant a nearly fourfold increase in the volume of goods transported around the world by shipping since 2000.

Why Does the Blue Acceleration Matter?

The ocean was once thought—even among prominent scientists—to be too vast to be changed by human activity. That view has been replaced by the uncomfortable recognition that not only can humans change the ocean but also that the current trajectory of human demands on the ocean simply isn't sustainable.

Consider the coast of Norway. The region is home to a multimillion-dollar oceanic oil and gas industry, aquaculture, popular cruises, busy shipping routes, and fisheries. All of these interests are vying for the same ocean space, and their demands are growing. A fivefold increase in the number of salmon grown by aquaculture is expected by 2050, while the region's tourism industry is predicted to welcome a fivefold increase in visitors by 2030. Meanwhile, vast offshore wind farms have been proposed off the southern tip of Norway.

The ocean is vast, but it's not limitless. This crowding in ocean space is not unique to Norway, and a densely populated ocean space runs the risk of conflict across industries. Escapee salmon from aquaculture have spread sea lice to wild populations, creating tensions with Norwegian fisheries. An industrial accident in the oil and gas industry could cause significant damage to local seafood and tourism as well as the seafood export market.

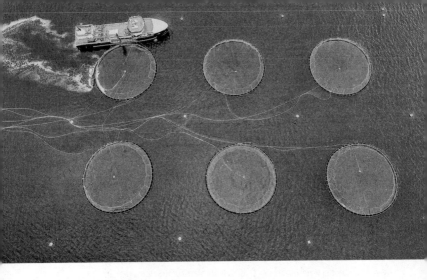

More fundamentally, the burden on ocean ecosystems is growing, and we simply don't know as much about these ecosystems as we would like. An ecologist once quipped that fishery management is the same as forestry management. Instead of trees, you're counting fish—except you can't see the fish, and they move.

Exploitation of the ocean has tended to precede exploration. One iconic example is the scaly-foot snail. This deep-sea mollusk was discovered in 1999 and by 2019 was on the Red List of endangered species of the International Union for Conservation of Nature. Why? As far as scientists can tell, the species is found in only three hydrothermal vent systems more than 2,400 meters below the Indian Ocean and covering less than 0.02 square kilometers. Today, two of the three vent systems fall within exploratory mining leases.

What Next?

Billionaires dreaming of space colonies can dream a little closer to home. Even as the Blue Acceleration consumes more

A salmon farm off the coast of Bergen, Norway.
MariusLtu/Getty Images

of the ocean's resources, this vast area is every bit as myste-
rious as outer space. The surfaces of Mars and the Moon have
been mapped in higher resolution than the seafloor.[3] Life in
the ocean has existed for two billion years longer than on land,
and an estimated 91 percent of marine species have not been
described by science.[4] Their genetic adaptations could help
scientists develop the antibiotics and medicines of tomorrow,
but they may disappear long before that's possible.

The timing is ripe for guiding the Blue Acceleration
toward a more sustainable and equitable trajectory. The
United Nations' Decade of Ocean Science for Sustainable
Development began in 2021,[5] a new international treaty on
ocean biodiversity is in its final stages of negotiation, and in
June 2022, governments, businesses, academics, and civil
society assembled for the United Nations Ocean Conference
in Lisbon.

Yet many basic questions remain. Who is driving the Blue
Acceleration? Who is benefiting from it? And who is being left
out or forgotten? These are all urgent questions, but perhaps
the most important and hardest question to answer of all is how
to create connections and engagement across all the interested
parties. Otherwise, the drivers of the Blue Acceleration will be
like the fish in the ecologist's analogy: constantly moving, invisible, and impossible to manage—before it is too late.

Notes

1.   Jean-Baptiste Jouffray, Robert Blasiak, Albert V. Norström, Henrik Öster-
     blom, and Magnus Nyström, "The Blue Acceleration: The Trajectory of

Human Expansion into the Ocean," *One Earth* 2, no. 1 (2020): 43–54, https://doi.org/10.1016/j.oneear.2019.12.016.

2. Robert Blasiak, Jean-Baptiste Jouffray, Colette C. C. Wabnitz, Emma Sundström, and Henrik Österblom, "Corporate Control and Global Governance of Marine Genetic Resources," *Science Advances* 4, no. 6 (2018): https://doi.org/10.1126/sciadv.aar5237.

3. "Mapping Our Planet, One Ocean at a Time," National Centers for Environmental Information, September 21, 2018, https://www.ncei.noaa.gov/news/mapping-our-planet-one-ocean-time.

4. Camilo Mora, Derek P. Tittensor, Sina Adl, Alastair G. B. Simpson, and Boris Worm, "How Many Species Are There on Earth and in the Ocean?," *PLOS Biology* 9, no. 8 (2011): https://doi.org/10.1371/journal.pbio.1001127.

5. "United Nations Decade of Ocean Science for Sustainable Development (2021–2030)," UNESCO, https://en.unesco.org/ocean-decade.

Chapter 32

Why Indigenous Knowledge Should Be an Essential Part of How We Govern the World's Oceans

MEG PARSONS and LARA TAYLOR*

OUR *MOANA* (OCEAN) is in a state of unprecedented eco-logical crisis. Multiple, cumulative impacts include pollution, sedimentation, overfishing, drilling, and climate change. All affect the health of both marine life and coastal communities. To reverse the decline and avoid reaching tipping points, we must adopt more holistic and integrated governance and management approaches.

---

*Independent researcher Payoshni Mitra contributed to this chapter.

Indigenous peoples have cared for their land and sea-scapes for generations by using traditional knowledge and practices. As our research on marine justice shows,[1] Indigenous peoples face ongoing challenges,[2] as they seek to assert their sovereignty and authority in marine spaces.[3] We don't need to wait for innovative Western science to take better care of the oceans. We have an opportunity to empower traditional and contemporary Indigenous forms of governance and management for the benefit of all people and the ecosystems we are part of.

Our research highlights alternative governance and management models to improve equity and justice for Indigenous peoples. These range from shared decision-making with governments (co-governance) to Indigenous peoples regaining control and reenacting Indigenous forms of marine governance and management.

Indigenous Environmental Stewardship

Throughout Oceania, Indigenous marine governance is experiencing a revival.[4] The long-term environmental stewardship of Indigenous peoples is documented around the globe.

In Fiji, customary marine tenure is institutionalized through the *qoliqoli* system. This defines customary fishing areas in which village chiefs are responsible for managing fishing rights and compliance. Coastal communities in Vanuatu continue to create and implement temporary marine protection zones (known as *tapu*) to allow stock in fisheries to recover. In Samoa, villages are able to establish and enforce local fisheries management.

In Aotearoa New Zealand, Māori environmental use and management is premised on the principle of *kaitiakitanga* (environmental guardianship) rather than unsustainable extraction of resources. Australian Aboriginal societies likewise use the term *caring for country* to refer to their ongoing and active guardianship of the lands, seas, air, freshwater, plants, animals, spirits, and ancestors.

From the Mountains to the Sea

These governance and management systems are based on Indigenous knowledge that connects places and cultures and emphasizes holistic approaches. The acknowledgment of interrelationships between human and nonhuman beings (plants, animals, forests, rivers, and oceans) is a common thread. So is an emphasis on reciprocity and respect toward all beings.

Coastal and island Indigenous groups have specific obligations to care for and protect their marine environments and to use them sustainably. An intergenerational thread is part of these ethical duties. It takes into account the lessons and experiences of ancestors and considers the needs of future generations of people, plants, animals, and other beings.

In contrast to Western ways of seeing the environment, the Australian Indigenous concept of country is not fragmented into different types of environment or scales of governance. Instead, land, air, water, and the sea are all linked. Likewise, for Māori, *Ki uta ki tai* (from the mountains to the sea) encapsulates a whole-of-landscape and -seascape view.

Sharing Knowledge across Generations

Māori hold deep relationships with their *rohe moana* (saltwater territory). These are increasingly recognized by laws that emphasize Indigenous Māori rights based on Te Tiriti o

Waitangi. One example is the Kaipara Moana Remediation Programme, the largest landscape project under way in Aotearoa. The Programme is a partnership between central and regional government agencies and Kaipara Uri (Ngā Maunga Whakahī o Kaipara, Te Rūnanga o Ngāti Whātua, and Te Uri o Hau—tribal descendants of the Kaipara Moana [harbor]).

The partners are embarking on a decade of restoration (estimated at AU$300 million) to address the impacts of historical land clearance and land use change within the catchment, which has caused elevated levels of sedimentation in waterways that ultimately deposits on the bed of Kaipara Moana. If this deposition is left unchecked, it could degrade Kaipara Moana beyond repair.

The Kaipara Moana Remediation Programme is intended to address this degradation through government funding, together with Kaipara Uri, who are using their *mātauranga* Māori (Māori Knowledge) alongside scientific knowledge to enact *kaitiakitanga* and ecosystem-based management. Together, those involved are co-developing new models of co-governance, planning, collaboration, and problem solving to achieve their vision to protect, restore, and enhance the *mauri* (life force) of the Kaipara Moana for present and future generations.[5]

Another co-management agreement is operating in Hawai'i between the community of Hā'ena (USA) and the Hawai'ian state government. The Hā'ena community operates an Indigenous fishing education program. Members of all ages camp together by the coast and learn where, what, and how to harvest and prepare marine products. In this way, Indigenous knowledge, with its emphasis on sustainable practices and environmental ethics, is transmitted across generations.

Indigenous knowledge, values, and relationships with our ocean can make significant contributions to marine governance. We can learn from Indigenous worldviews that emphasize connectivity among all things. There are many similarities between ecosystem-based and Indigenous knowledge management systems.

We need to do more to recognize and empower Indigenous knowledge and ways of governing marine spaces.[6] This could include new laws, institutions, and initiatives that allow Indigenous groups to exercise their self-determination rights and draw on different types of knowledge to maintain sustainable seas.

Notes

1. Meg Parsons, Lara Taylor, and Roa Crease, "Indigenous Environmental Justice within Marine Ecosystems: A Systematic Review of the Literature on Indigenous Peoples' Involvement in Marine Governance and Management," *Sustainability* 13, no. 8 (2021): 4217, https://doi.org/10.3390/su13084217.
2. Merle Sowman and Jackie Sunde, "Social Impacts of Marine Protected Areas in South Africa on Coastal Fishing Communities," *Ocean & Coastal Management* 157 (2018): 168–79, https://doi.org/10.1016/j.ocecoaman.2018.02.013.
3. Laurie Richmond and Dawn Kotowicz, "Equity and Access in Marine Protected Areas: The History and Future of 'Traditional Indigenous Fishing' in the Marianas Trench Marine National Monument," *Applied Geography* 59 (2015): 117–24, https://doi.org/10.1016/j.apgeog.2014.11.007.
4. Heather L. McMillen, Tamara Ticktin, Alan Friedlander, Stacy D. Jupiter, Randolph Thaman, John Campbell, Joeli Veitayaki, et al., "Small Islands, Valuable Insights: Systems of Customary Resource Use and Resilience to Climate Change in the Pacific," *Ecology and Society* 19, no. 4 (2014): https://doi.org/10.5751/es-06937-190444.
5. Leane Makey and Shaun Awatere, "*He Mahere Pāhekoheko mō Kaipara Moana*–Integrated Ecosystem-Based Management for Kaipara Harbour, Aotearoa New Zealand," *Society & Natural Resources* 31, no. 12 (2018): 1400–1418, https://doi.org/10.1080/08941920.2018.1484972.
6. Chantel E. Collier, "Enabling Conditions for Community-Based Comanagement of Marine Protected Areas in the United States," *Marine Policy* 122 (2020): 104244, https://doi.org/10.1016/j.marpol.2020.104244.

Contributors

Roger Bales, Distinguished Professor of Engineering at the University of California, Merced, and Adjunct Professor at the University of California, Berkeley, focuses his teaching and research on building the knowledge base needed to adapt California's water supplies, forests, and economy to climate warming. His scholarship includes more than 200 peer-reviewed articles and many more reports in publicly accessible media. He works with research colleagues, leaders in state agencies, elected officials, land managers, water leaders, nongovernmental organizations, and other key decision makers on developing climate solutions for California.

Kevin M. Befus is an Assistant Professor at the University of Arkansas and a hydrogeologist who studies interconnected hydrologic systems, with a focus on groundwater. His research addresses problems of groundwater sustainability and groundwater dynamics. He employs numerical modeling to test processes and perform sensitivity analyses. He occasionally has the opportunity to conduct observational studies with fieldwork.

Robert Blasiak is a researcher at the Stockholm Resilience Centre (Stockholm University), where he focuses on the sustainable management of ocean resources and ocean stewardship. His recent work has dealt with issues surrounding the conservation and equitable use of marine genetic resources. He was previously a Senior Research Fellow with the Nippon Foundation NEREUS Program, as well as a Visiting Researcher at the University of Tokyo and the United Nations University.

Ellen Bruno is an Assistant Professor of Cooperative Extension in the Department of Agricultural & Resource Economics at the University of California, Berkeley. Dr. Bruno received her PhD and MS degrees in Agricultural and Resource Economics from the University of California, Davis, and her BS degree in Management Science from the University of California, San Diego. As an extension economist, she conducts research and outreach that focus on policy issues relevant to California's agriculture and natural resources. Her recent research, which has appeared in journals such as the *American Journal of Agricultural Economics* and the *Journal of Public Economics*, considers the potential for and effectiveness of water-related policies.

Bethany Caruso is an Assistant Professor in the Hubert Department of Global Health in the Rollins School of Public Health at Emory University. She is a social and behavioral scientist with over a decade of experience in water, sanitation, and hygiene (WASH) sectoral research. She employs mixed-methods approaches to understanding how compromised WASH conditions impact physical and mental health, behavior, education, and empowerment, with a specific focus on girls and women throughout their life course.

Sebastien Chastin is a Professor of Health Behavior Dynamics of people, places, and systems at Glasgow Caledonian University and Ghent University. He has a background in physics and rehabilitation sciences. He is a fellow of the Royal Statistical Society. Previously he had a post at the British Antarctic Survey, Oxford and Edinburgh Universities. His research focuses on the dynamics of health behavior in relation to aging, places, and systems. Understanding why, when, and how people decide to move or not move is crucial to promoting healthy movement behavior and aging.

Craig E. Colten has studied the geography of hazards for over 30 years. He launched his career by investigating practices of hazardous wastes management in the era before the Environmental Protection Agency, particularly seeking to document long-forgotten disposal sites. More recently, his attention turned toward urban hazards, community resilience, and adaption to environmental change. Assisted by colleagues and students, he has considered the specific practices used at the community and family level that have enabled people to rebound from devastating events like hurricanes, giant oil spills, and river floods. His work can be found in the books *An Unnatural Metropolis*; *Perilous Place, Powerful Storms*; *Southern Waters*; and *State of Disaster*.

Joseph Cook focuses his research on water and sanitation policy in low-income countries, water resources economics and policy, and non-market valuation. He has an MS and PhD from the University of North Carolina at Chapel Hill and a BS from Cornell University. His research has appeared in outlets such as the *Journal of the Association of Environmental and Resource Economists*, *Environmental and Resource Economics*, *Water Resources Research*, the *Journal of Policy Analysis and Management*, and *World Development*.

Michelle DiBenedetto conducts research at the intersection of environmental fluid mechanics and particle-laden flows. A large part of her work is motivated by a desire to understand microplastics pollution in the ocean and problems in biological fluid mechanics. She primarily uses laboratory experiments and mathematical modeling in her research. She received her BS in Environmental Engineering from Cornell University in 2014 and a PhD in Civil and Environmental Engineering from Stanford University in 2019. After completing a postdoctoral position at Woods Hole Oceanographic Institution, she joined the faculty of mechanical engineering at the University of Washington in 2020.

David Feldman specializes in water resources management and policy, global climate change policy, ethics and environmental decisions, adaptive management, and sustainable development. His current research is focused on green infrastructure and urban water policy, resolution of trans-boundary dispute over water, flood risk communication, the water-energy policy nexus, and the challenges to achieving institutional reform that promotes equity in water management.

Farshid Felfelani is a Project Scientist in the Research Applications Laboratory at the National Center for Atmospheric Research. His research seeks to advance our understanding of hydrology-human-climate interactions in response to the combined effects of human activities and climate change by improving the representation of coupled natural-human hydrologic systems in land models. In particular, he uses models to address the global issues associated with the quantity and quality of terrestrial water systems under climate change and socioeconomic impacts. He has published in high-impact journals including *Nature Climate Change*, *Water Resources Research*, *Geophysical Research Letters*, the *Journal of Hydrology*, and *Hydrological Sciences Journal*.

Gabriel Filippelli is a Chancellor's Professor of Earth Sciences at Indiana University–Purdue University Indianapolis. He is a biogeochemist focusing on the flow and cycling of elements and chemicals in the environment. This focus includes his work on pollutant distribution and exposure for human populations, as well as ways to engage communities in reducing their own exposure. He is also the Executive Director of the Indiana University Environmental Resilience Institute.

Michail Georgiou is currently a PhD researcher at the School of Health and Life Sciences of Glasgow Caledonian University, Scotland, United Kingdom. He has a background in economics and environmental sciences. His present research project, titled "Impact of Blue Spaces on Health and Health Equality," focuses on mechanisms in the relationship between waterways and public health. He is taking an epidemiological, physiological, and economic approach.

Burke W. Griggs is a Professor of Law at Washburn University School of Law. He teaches property law to first-year students and natural resources law to upper-division students. His scholarship explores the historical, technical, and cultural aspects of American natural resources law, especially water law. He has authored numerous scholarly articles and is the coauthor of standard casebooks on both water law and oil and gas law. He is affiliated with the Woods Institute for the Environment and the Bill Lane Center for the American West, both at Stanford University, and serves as a Trustee of the Foundation for Natural Resources and Energy Law.

Gary Griggs is a Professor of Earth Sciences at the University of California, Santa Cruz. His research is focused on the coastal zone and ranges from coastal evolution and development, through shoreline processes, coastal hazards and coastal engineering, and sea level rise. Professor Griggs has published over 200 articles in scientific journals and written or cowritten 13 books.

Drew Gronewold has research interests in hydrological modeling, with a focus on propagating uncertainty and variability into model-based decisions in water resources management. His specific research areas include predicting runoff in ungauged basins, monitoring and understanding water quality dynamics in coastal areas, and incorporating probability theory and Bayesian statistics into watershed-scale datasets and forecasting tools. He holds adjunct appointments in the University of Michigan's Department of Civil and Environmental Engineering and its Department of Earth and Environmental Science. Prior to his appointment in the university's School for Environment and Sustainability, he worked as a hydrologist and physical scientist in the Great Lakes Environmental Research Laboratory of the National Oceanic and Atmospheric Administration.

Marissa Grunes is a science writer focused on the earth sciences, environmental history, and American literature and culture. She holds a PhD in English Literature from Harvard University.

Danielle Hare is a hydrologist with expertise in the process-based mechanics of surface water and groundwater connectivity and bio-geochemical processing within these environments. Her research has explored how groundwater connectivity affects wetland restoration, using temperature as a tool to determine groundwater influence, and how groundwater influence on stream discharge determines ecological resilience to climate change. She previously worked as an environmental consultant and is currently in a PhD program at the University of Connecticut, working with Dr. Ashley Helton to understand the effects of temperature increase on carbon dynamics within stream networks.

Brian Haus is the Chair of and a Professor in the Department of Ocean Sciences in the University of Miami's Rosenstiel School of Marine and Atmospheric Science. He is currently conducting experimental studies to quantify coastal land-air-sea interaction to improve environmental models and to develop innovative hybrid natural and engineered structures for coastal resilience. His focus is on fundamental processes that occur at the interfaces between the fluid atmosphere and ocean and solid surfaces. Understanding the transfers of heat, moisture, momentum, and energy that occur across these surfaces can lead to improved climate modeling, hurricane and wave forecasting, safer marine activities, and more resilient coastal communities.

Daniel Johnson is an Associate Professor of Tree Physiology and Forest Ecology in the Warnell School of Forestry and Natural Resources at the University of Georgia. His areas of specialty include forest biology, tree physiology, and ecology. He is particularly interested in plants' water transport and drought tolerance and in tree function in a changing climate.

Carol Kwiatkowski is an Adjunct Assistant Professor of Biological Sciences at North Carolina State University and a Senior Associate at the Green Science Policy Institute. Over the past two decades, she has accumulated a unique breadth of experience in researching the interaction of humans and their environment. This experience includes positions in academia, conducting primary research on human behavior and disease transmission, and positions in the nonprofit sector, analyzing and reviewing environmental chemicals and their impacts on health. Kwiatkowski enjoys making scientific research approachable and valuable to nonscientists. She has produced written materials and delivered oral presentations—for academics, policy makers, advocates, and the public—in which she shares science to support health-protective decision-making.

Rosalyn R. LaPier is a Professor of Environmental Studies at the University of Montana. She is an award-winning Indigenous writer and ethnobotanist with a BA in Physics and a PhD in Environmental History. She studies the intersection of traditional ecological knowledge learned from elders and the academic study of environmental history and religion. She is an enrolled member of the Blackfeet Tribe in Montana and Métis.

Katharine Mach is a Professor in the University of Miami's Rosenstiel School of Marine and Atmospheric Science and a faculty scholar at the University of Miami's Abess Center. Her research assesses climate change risks and response options to address increased flooding, extreme heat, wildfires, and other hazards. Mach is the 2020 recipient of the Piers Sellers Prize for a world-leading contribution to solution-focused climate research. Mach is a chapter lead for the US Fifth National Climate Assessment and was a lead author of the Sixth Assessment Report from the Intergovernmental Panel on Climate Change.

Amahia Mallea, PhD, is an environmental historian and author of the book *A River in the City of Fountains: An Environmental History of Kansas City and the Missouri River* (University Press of Kansas, 2018). She is an Associate Professor of History at Drake University and is interested in the relationship between American societies and their lands and resources. Subjects of interest include cities, rivers, and agriculture.

Daniel Craig McCool conducts research on water resources and Indian voting and water rights. His books include *River Republic: The Fall and Rise of America's Rivers* (Columbia University Press, 2012) and *Native Waters: Contemporary Indian Water Settlements and the Second Treaty Era* (University of Arizona Press, 2002). His most recent book, coedited, is *Vision and Place: John Wesley Powell and Reimagining the West* (University of California Press, 2020).

Jacob A. Miller-Klugesherz is a sociology PhD student at Kansas State University with a research traineeship sponsored by the National Science Foundation. He researches social and political barriers to regenerative agriculture adoption, community and personal well-being, absentee ownership, moral foundations, and ecospheric rhetoric. His research has found homes in the journals *Frontiers in Water*, *Kybernetes*, and more. He is a sixth generation Kansan, born and raised.

Nobuhito Mori is a Vice-Director and Professor at the Disaster Prevention Research Institute, Kyoto University. He has also been an Honorary Professor at Swansea University, United Kingdom, since 2020 and a Visiting Professor at Yokohama National University, Japan, in 2021, and a Visiting Professor at Hiroshima University, Japan, in 2022. His areas of interest include air-sea interface physics, climate change, and the dynamics of wind waves, long waves, and tsunamis. He also has expertise in geophysical and environmental fluid mechanics and in tropical cyclones and related disasters.

Thomas Mortlock is a Senior Analyst at Aon and an Adjunct Fellow at Macquarie University with a background in climate risk and coastal and flood modeling. He has worked for over 14 years in the fields of climate science, catastrophe risk and insurance, and coastal and flood modeling. He has supported corporations and government in understanding the materiality of climate risk for their business and operations. He is a Chartered Engineer with a PhD in Environmental Science. He has authored more than 20 peer-reviewed papers in coastal and climate risk. He currently sits on the Australia Pacific Climate Partnership Expert Panel.

Suzanne O'Connell uses marine sediment cores (libraries of earth history) to study oceans' past and climate. She has sailed on nine scientific ocean-drilling expeditions. The most recent was Expedition 382, in spring 2019, to Iceberg Alley. She will sail to the North Atlantic in summer 2023. Diversity in geoscience is her passion. She coedited a book on gender equality and helped start the On To the Future program at the Geological Society of America to broaden participation in the geosciences. More than half of her research students are from underrepresented groups. She was the Director of Wesleyan University's McNair Program for five years.

Itxaso Odériz graduated as a doctor from the National Autonomous University of Mexico and a civil engineer from Cantabria University. Currently she works as a Postdoctoral Researcher at the Environmental Hydraulics Institute of the Universidad de Cantabria (IHCantabria). She has proposed global scenarios of natural variability and climate change and examined their role in coastal hazards using big-data approaches. She has also worked on coastal adaptation strategies using green solutions, such as dune vegetation and coral reef–mimicking infrastructure. Her research links large-scale climate patterns with local coastal processes to face climate change. Her research in climate change won the Miguel Urquijo Award and the first award for Best Research in Climate Change PINCC 2021.

Joseph D. Ortiz is a Professor in the Department of Earth Sciences at Kent State University. With training in aquatic biology and oceanography, he works at the interface among unraveling climate mysteries with science, exploring the relationship between sedimentary strata, improving water quality using remote-sensing techniques, and evaluating the transition to a sustainable energy system. His primary research interests are in paleoclimate and environmental remote sensing.

Meg Parsons is an Aotearoa New Zealand historical geographer, of both Māori (Indigenous) and Pākehā settler (European) heritage, whose work examines the intersecting impacts of colonization on Indigenous and settler societies, with a particular focus on the resulting social and environmental changes for Indigenous peoples in Australia, Aotearoa New Zealand, and the Pacific. Parsons's current research examines the different ways in which Indigenous people are seeking to reassert their Indigenous knowledges, governance, and environmental management to address multiple forms of injustice, as well as efforts by Indigenous and non-Indigenous groups to be included in climate adaptation planning and actions in Oceania.

Raquel Partelli-Feltrin is currently a Postdoctoral Scholar in the Department of Botany at the University of British Columbia, Canada. Her research is focused on tree physiology and fire science. She is particularly interested in how changes in fire behavior can impact the physiological responses of trees such as their water transport and carbon pools.

Yadu Pokhrel is an Associate Professor in the Department of Civil and Environmental Engineering at Michigan State University. His research focuses on improving our understanding of changes in water, food, and energy systems in response to the combined effects of human activities and climate change. His work is transdisciplinary and engages diverse stakeholders and is funded primarily by the National Science Foundation (NSF) and the National Aeronautics and Space Administration. He has published 80-plus peer-reviewed articles, including 10-plus in the journals *Nature* and *Science*. He is a recipient of several awards, including the NSF CAREER Award and US Fulbright Award. He serves as an Associate Editor for *Water Resources Research* and the *Journal of Hydrology*.

Manzoor Qadir is the Deputy Director of the United Nations University's Institute for Water, Environment and Health, based in Canada. His expertise covers water-related sustainable development through contributions to policy, institutional and biophysical aspects of unconventional water resources, water recycling and safe reuse, water quality and environmental health, and water and food security in a changing climate. He previously held professional positions at the International Center for Agricultural Research in the Dry Areas and the International Water Management Institute; was an Alexander von Humboldt Fellow and Visiting Professor at Justus-Liebig University, Germany; and was an Associate Professor at the University of Agriculture, Pakistan.

Julie Reimer, PhD, is an ocean scientist and advocate. Her research explores marine spatial planning and area-based ocean management as a pathway toward global conservation and sustainability goals. She holds a PhD in Geography, a Master of Marine Management, and a Bachelor of Science in Biology and brings this interdisciplinary training to her work. She has been recognized as a leader in organizational governance and science-based conservation advocacy and, in 2021, was named a Top 30 Under 30 Sustainability Leader in Canada.

Landolf Rhode-Barbarigos is an Assistant Professor in the Department of Civil and Architectural Engineering at the University of Miami. His research focuses on structural morphology and morphogenesis. Structural morphology is the study of the relation between a structure and its function, form, material, and forces, while structural morphogenesis, in an analogy with biology, refers to the processes that control the organized spatial distribution of material and modules in a structural system. He explores these concepts to design tomorrow's structures with applications varying from marine and coastal structures to building and infrastructure systems to space structures.

Richard B. "Ricky" Rood is a Professor at the University of Michigan in the Department of Climate and Space Sciences and Engineering and is also appointed in the School for Environment and Sustainability. He is part of the Core Team of the Great Lakes Integrated Sciences and Assessments Center. Professor Rood has made research contributions to several fields. His numerical algorithms are used in climate models, weather-forecast models, and atmospheric chemistry models. He has also been a leader in developing merged model–observation datasets to study chemistry and climate. His current research concerns the usability of climate science in adaptation planning.

Asher Rosinger studies human biology as an Assistant Professor of Biobehavioral Health and Anthropology at Pennsylvania State University. He founded and directs the Water, Health, and Nutrition Laboratory. The laboratory aims to understand cross-cultural variation in how humans meet their water needs across distinct ecological contexts and how this relates to both acute and long-term health, hydration status, nutritional status, stress, and disease risk.

Matthew R. Sanderson is Randall C. Hill Distinguished Professor of Sociology, Anthropology, and Social Work and a Professor of Geography and Geospatial Sciences at Kansas State University. His research empirically examines population and environment as aspects of development. Most of his work is international in scope and longitudinal and comparative in design. He leads courses on international development and social change, environment and society, international migration, rural development, and principles of social science.

Heidi Schweizer is an applied economist whose research is at the intersection of commodity marketing, transportation, and supply chain management. She holds a PhD in Agricultural and Resource Economics from the University of California, Davis. She is currently an Assistant Professor and Cooperative Extension Specialist at North Carolina State University, where she teaches a class on futures and options markets and addresses freight market issues faced by the agricultural community in North Carolina.

Alan Seltzer is a geochemist who studies the chemistry of gasses in nature to trace physical and biogeochemical interactions among the atmosphere, ocean, hydrosphere, cryosphere, and solid Earth. His research is primarily directed at climate–relevant questions, including how atmospheric gasses are taken up by the deep ocean and how hydroclimate on land has evolved since the last ice age. As an Assistant Scientist at Woods Hole Oceanographic Institution (Woods Hole, Massachusetts), he runs a lab that specializes in measuring small signals of inert gas isotopes dissolved in seawater and groundwater and in interpreting these signals with physical models.

A. R. Siders is an Assistant Professor at the Disaster Research Center, the Biden School of Public Policy and Administration, and the Department of Geography and Spatial Sciences at the University of Delaware. Siders's research is focused on climate change adaptation policy, seeking to understand how adaptation decisions are made and how those decisions affect social justice and risk reduction outcomes. Siders is a codirector of the Gerard J. Mangone Climate Change Science and Policy Hub.

Rodolfo Silva is a Full Researcher at the Institute of Engineering of the National University of Mexico (UNAM). His main area of research is coastal engineering, including the design of maritime works, coastal morphodynamics, oceanographic risk, marine energy, and the restoration of ecosystems. Since 1995 he has headed the Coastal and Oceanographic Group at UNAM and the Mexican Centre for Ocean Renewable Energies (CEMIE-Océano). He has produced more than 110 engineering technical reports for national and international bodies. As author or coauthor, he has more than 570 refereed papers or chapters published in journals, books, and conference proceedings.

Vladimir Smakhtin is the Director of the United Nations University Institute for Water, Environment and Health, based in Canada. He has over 35 years of policy-relevant research experience in global water resources in international development. He holds a PhD in Hydrology and Water Resources from the Russian Academy of Sciences and has worked in Russia, in South Africa, and at the International Water Management Institute headquartered in Sri Lanka. His expertise spreads across environmental water management, river basin modeling, water-related disaster risk management, water scarcity assessment, and water storage planning.

Bruce Sutherland studies the dynamics of the atmosphere, ocean, and estuaries through laboratory experiments, numerical simulations, and mathematical theory. His research focuses on waves, currents, and convection and how these influence particle transport, sedimentation, and resuspension. After receiving his PhD in 1994 at the University of

Toronto, he was a Research Associate at the University of Cambridge until his appointment at the University of Alberta in 1997. He is presently a Professor jointly appointed in the Departments of Physics and of Earth and Atmospheric Sciences. In addition to his research and teaching, he is an Associate Editor for *Physical Review Fluids*.

Lara Taylor is of Māori (Ngāti Tahu Ngāti Whaoa, Ngāti Kahungunu, Ngai Tahu), Dutch, and Pākehā descent. She works as a Kairangahau Māori (Indigenous researcher) to enable and empower Indigenous tribes, sub-tribes, families, and people through improved environmental planning, policy, and legislation. She is a Trustee for one of her tribal land blocks and works for her people of Ngāti Tahu regarding their geothermal resource. She is a very proud māmā of three.

Emily Ury is a Postdoctoral Fellow in the Department of Earth and Environmental Science at the University of Waterloo in Ontario. Her research focuses on wetlands, their response to global changes, and how to conserve and restore these valuable ecosystems. Emily received her PhD in Ecology from Duke University in 2021.

Ton van den Bremer is an Associate Professor in Fluid Mechanics at TU Delft (Delft University of Technology), a Senior Research Fellow and Royal Academy of Engineering Research Fellow at the University of Oxford, and a Visiting Professor at the University of Edinburgh. He is Principal Investigator for a research project titled "Cleaning the Ocean: Understanding Transport of Plastic Pollution by Waves."

Andrew J. Whelton has more than 20 years of experience uncovering and addressing problems at the interface of infrastructure materials, the environment, and public health. His team's discoveries have positively changed how government agencies, water utilities, nonprofit organizations, health departments, state legislatures, and building owners approach their responsibilities. Before joining Purdue University, he worked for the University of South Alabama, the National Institute of Standards and Technology, Virginia Tech University, the US Army, and private engineering consulting firms. Hallmarks of his work are directly engaging with communities at risk and making discoveries publicly accessible. He earned a BS, MS, and PhD in Civil and Engineering from Virginia Tech.

Index